THE PSYCHOLOGY OF EDUCATION AND INSTRUCTION

A series of volumes edited by:
Robert Glaser and **Lauren Resnick**

LINGUISTIC and CULTURAL INFLUENCES on LEARNING MATHEMATICS

Edited by

Rodney R. Cocking
National Institute of Mental Health

Jose P. Mestre
University of Massachusetts

LEA

LAWRENCE ERLBAUM ASSOCIATES, PUBLISHERS

1988   Hillsdale, New Jersey   Hove and London

The image on the cover is a perturbed genus-4 example of the Hoffman-Meeks embedded minimal surface discovered by David A. Hoffman and William H. Meeks. The image was generated by James T. Hoffman using a Ridge 32/110 computer, a Raster Technologies Model One/380 graphics controller and an Apple Laserwriter. This work was performed at the University of Massachusetts at Amherst.

Lawrence Erlbaum Associates, Inc., Publishers
365 Broadway
Hillsdale, New Jersey 07642

Library of Congress Cataloging in Publication Data

Linguistic and cultural influences on learning
   mathematics.

   Includes indexes.
   1. Mathematics—Study and teaching.   2. Language and
education.   I. Cocking, Rodney R.   II. Mestre, Jose P.
QA11.L718   1988   510′.7   87-22285
ISBN 0-89859-876-1

Printed in the United States of America
10   9   8   7   6   5   4   3

Contents

Contributors

Manon P. Charbonneau
University of New Mexico

Susan Chipman
Office of Naval Research

Rodney R. Cocking
National Institute of Mental Health

JoAnn Crandall
Center for Applied Linguistics

Theresa Corasaniti Dale
Center for Applied Linguistics

Edward A. DeAvila
DeAvila, Duncan & Associates

Richard P. Durán
University of California–Santa Barbara

Vera John-Steiner
University of New Mexico

William L. Leap
The American University

Patricia MacCorquodale
University of Arizona

Jose P. Mestre
University of Massachusetts

Anne M. Milne
Decision Resources Corporation

David E. Myers
Decision Resources Corporation

Nancy Rhodes
Center for Applied Linguistics

Geoffrey B. Saxe
University of California–Los Angeles

Ellin Kofsky Scholnick
University of Maryland

George Spanos
Center for Applied Linguistics

Joan B. Stone
*University of Rochester & Rochester
Institute of Technology*

Joseph H. Suina
University of New Mexico

Kenneth J. Travers
University of Illinois

Sau-Lim Tsang
ARC Associates

Foreword

It is a pleasure to write the Foreword for this important volume, *Linguistic and Cultural Influences on Learning Mathematics*. Over the past 15 years or so, research on the learning of mathematics has made important advances by focusing mainly on the "cognitive" aspects of the problem. Thus, we have learned that even 3-year-olds possess rudimentary notions of more and less; that school children's errors often result from "buggy" procedures; and that common strategies of mental addition are used in different cultures.

The Cocking and Mestre volume performs a valuable service by calling our attention to broader aspects of mathematics learning. *Linguistic and Cultural Influences on Learning Mathematics* brings the study of mathematics learning into the real world of schools, culture, and bilingualism. (To get a feel for the personal meaning of these issues you might begin this volume at the end, with Suina's eloquent Epilogue, "And Then I Went to School.") The volume's diverse chapters, drawing on many scholarly disciplines, explore these and other factors that make the learning of mathematics so difficult in American schools.

A basic theme of this volume is that the understanding of mathematics performance in the school requires much more than cognitive research alone. Although cognitive science has made an important contribution in showing that children from different cultures possess the potential for at least basic mathematics learning, we need to understand why that potential is seldom realized. Why is it that schooling is so often an unrewarding and unpleasant experience for so many of our children? To understand the failure of the educational system, we need to explore a host of problems usually slighted in cognitive research—problems of bilingualism, culture, class, gender, affect, learning style, motivation, the availability of sound educational opportunities, teacher competence, the implemented curriculum (as contrasted to the intended curriculum), and the like.

These issues are complex and messy, far harder to deal with than say the question of whether children's addition involves "counting on" or "counting all." But, if we are to provide genuinely equal access to a meaningful education, these are the problems that must be dealt with. A realistic and comprehensive understanding of learning in the real world of flawed education, economic inequality, and squandered intellectual potential must grapple with the issues raised in this volume. We are therefore indebted to the authors of the many fine chapters in this book for their efforts in raising fundamental questions of mathematics performance and in shedding light on them.

Herbert P. Ginsburg
Teachers College,
Columbia University

LINGUISTIC and CULTURAL INFLUENCES on LEARNING MATHEMATICS

Chapter 1

Introduction:
Considerations
of Language Mediators
of Mathematics Learning

Rodney R. Cocking
National Institute of Mental Health
Jose P. Mestre
University of Massachusetts

Classical language studies have long been regarded as important components of good academic training, especially during elementary and high school years. This philosophy guided individuals like John Mill in how they educated their own children, as well as providing the cornerstones of many early educational institutions (e.g., Boston Latin). The educational philosophy cast language in a central role for at least three reasons: (a) Language studies involve *discipline* in learning formal structures, thus promoting the development of structural organization to thought expression; (b) language studies provide an *historical perspective* that has implications both for language (knowledge of word classes and word origins) and for historical time perspective on the evolution of academic content; and (c) finally, language studies were thought to be important because, by learning the language of the original literature, learners had direct *access* to others' thinking. Thus, discipline, historical perspective, and access to knowledge were guiding rationales for making language studies a cornerstone to education.

Of these three reasons access, particularly as it pertains to educational opportunity, continues to enter into current debates over the role of language in learning. Battle lines have been drawn over political issues that are fueled by a plethora of legislation aimed at the access question. The controversy is over who deserves the largest percentage of a limited pool of school funds.

A brief look at the motivation behind access legislation will illustrate that none of the raging debates have anything to do directly with learning, understanding, or achievement. However, the effect of such debates greatly promotes or limits access to information and for this reason the policy debates are important to consider in the present context of language studies and school achievement.

The Civil Rights decision of 1974 stated that everyone deserves *equal access* to an education. The definition of *equal access* was unstated, but language skills were seen by many educators as limiting factors to some students' ability to benefit fully from classroom instruction. The causes of the language limitations were varied and the remedies provided by numerous pieces of legislation that were drafted to ensure equal access were also varied. Thus, although equal access was the overarching Civil Rights concern, how to apply the "equal access" criterion in the case of language skills generated internal debates among educators. Who deserves special language programs and what constitutes a language deficit drew, and continues to draw, the major attention in debates over the role of language skills in academic success. Language limitations have been defined legislatively by the following: *Title VII*—Elementary and Secondary Education Act that contained a bilingual education provision; *Title IV* of the Indian Education Act recognizes that students may need assistance in improving school language skills that differ from their tribal languages; Headstart antipoverty programs of *Title I* (Chapter I) recognize that language skills training may be necessary for a wide segment of the population in preparing them for later academic work; the *Indo-Chinese Refugee Act* contains language assistance provisions for non-native speakers of English for helping students enter public schools; and *P.L. 94.142*—Education for All Handicapped specifies that there is a wide variety of language performance areas that can constitute handicapping conditions and that can limit students' access to education.

These pieces of legislation recognize that students with below-norm language skills may be at risk educationally. This assumption is borne out of national educational statistics on who are underachievers in school. However, the relationship between language skills and academic performance is not a perfect one: some academic areas are affected more than others, some language skills appear to be more directly responsible for school success than others, and some language-minority groups are more adversely affected than others by the presence of more than one language in their environments. One academic area of current interest that appears to have a large language-related component is mathematics. The relationship between language skills and mathematics is not well understood, even though the same low math achievement statistics among certain groups of learners are reported year after year.

The legislative-related controversies and the educational statistics, like the older educational philosophies, continue to cast language in a central learning role. The purpose of this volume was to draw upon a wide array of experts from diverse fields to look at this important educational research problem. The intent was to pool the knowledge and methodologies of researchers in cognitive development, education, language studies, and mathematics to look at the problem of poor mathematics skills among language minority learners. The available literature indicated clearly that we were not dealing solely with mathematics learning or solely with language issues. Rather, the consensus was that cognitive processes of learning, comprehension, and symbolic thinking were all involved. That is, the topic was one of cognitive processes, rather than one singly determined by math or language skills.

RESEARCH QUESTIONS AND APPROACHES

We began by asking a series of questions that had answers or potential answers in differing sources. First, we asked what we know about the mathematics-language relationship that is potentially helpful in improving achievement for students experiencing difficulties in mathematics. Second, we asked where the research gaps exist pertaining to what we need to know about mathematical understanding and mathematics learning among language minority students. Finally, we asked what data banks and data sources exist that might be mined to address questions without collecting new data or that might be useful for pilot work before formulating new, expensive data-gathering projects.

A two-pronged approach was devised to look at these questions. One prong consisted of formulating the important research questions that bear on the topic. This activity included asking experts to write conceptual papers representing a variety of viewpoints, and tapping recent research projects for analyses that would address some of the relevant questions. Because we were dealing with a multiply determined problem, several substantive areas were identified as important in analyzing the problem. Specifically, the following areas were included: knowledge of large data bases on mathematics achievement, how language facilitates and/or limits one's ability to access one's knowledge, school performance among low-socioeconomic status students, the interaction of language and conceptual development, development of early number concepts, issues of intellectual development, and cultural background perspectives on development.

The second prong of the activity was simultaneously to identify and document existing data banks that might be appropriate for conducting secondary analyses. Secondary analyses, it was thought, could be useful in formulating and answering new research questions, as well as in fleshing-out a research model. The larger tasks of identifying and documenting available data bases had to be accomplished before secondary analyses could be performed. Well-known national data bases were easy to survey for what they had to offer, for the adequacy of documentation, and the overlap of previous reports with our current objectives. Federally funded research projects and state-level data bases were eventually identified but were usually problematic because of an inability to compare data across sites for pooling samples, because of the lack of information regarding the sample sizes themselves, or because of inadequate documentation. Data from individual researchers were identifiable only to the extent that projects were well-known through the published literature or by personal contact, and these data bases were small in scope and designed for such specific purposes as to contain few of the variables that overlapped across projects or with larger data bases. From over 160 data bases that we identified, only 18 potentially could be used in the secondary analyses.

Table 1.1 is a summary matrix of the relevant characteristics of data bases considered for possible inclusion in the secondary analyses of mathematics achievement of language minority students. A number of these data bases had potential for usefulness, but some general points had to be considered.

Generally, independent variables such as the student's home language and language proficiency, as well as parental socioeconomic status (SES) were inconsistent across various data sets. Sometimes the non-English language variable was the language the student usually uses and sometimes it was the language used by others in the home. The extent to which various languages were used was often missing and definitions varied. Presence of a non-English language for children was variously reported by students, teachers, or parents. Proficiency data were almost totally lacking across all data sets. Socioeconomic indicators were almost as inconsistent as language indicators: Some data bases used parental educational attainment or incomes, whereas others used the types of indicators that are often used by school districts to allocate various sources of funds (eligibility for free or reduced-price lunches, for example). These kinds of variability made it difficult to equate data bases.

It should be noted that there was also a fair degree of variability across the mathematics achievement variable. Some data bases contained standardized test scores whereas others used criterion-referenced tests. Further, not all data bases had scores for both mathematics and reading achievement, so comparisons across subject matter fields was often difficult.

Data from states or large school districts often presented two types of problems: First, many state data bases are available only in aggregate form—for school districts or for entire schools. Second, access to state data bases is often difficult and time-consuming. States have various procedures that are designed to protect the confidentiality of their data and one may have to gain approval from various levels and pay to have personal identifiers stripped from the records. All of this takes time and increases the cost of secondary analyses.

Finally, the data bases that offer greatest utility in measuring the language and SES variables (i.e., the nationally representative studies) are generally lacking some of the finer grained educational process variables, such as teacher characteristics and time-on-task. Smaller studies that had in-depth measures of mathematics learning processes and problems often contained no measures of mathematics achievement.

Despite these problems, several large data bases were appropriate and a number of important questions could be addressed with a sample size large enough to yield meaningful information. The chapter by Myers and Milne summarizes several sets of secondary analyses as well as a research model that guided much of our thinking.

The chapters in the present volume represent a convergence of the two prongs outlined. Various analyses of data contributed to revisions and rethinking of concept issues; conceptual papers that were circulated among the chapter authors influenced the kinds of analyses that were suggested for the secondary analyses and what writers ultimately identified as areas needing more and/or better information.

In addition to these fine-grained analyses and attempts to tease apart the issues, exchanges among the contributors of this volume often involved an interplay between a search for detail and attempts to get a clearer focus on the larger picture of the educational concerns. The outcome of such interplay and dynamics among the contributors represented here is an emphasis on some subtle issues that, we think, provide valuable insight into how research should proceed. For instance, it was agreed among the writers that both mathematics and language were symbol systems, and therefore symbolic thinking was a relevant cross-over area to consider. Likewise, some writers argued that math itself is a language and therefore language-learning issues were important in thinking about the various strands of math representation. One of the contributors to this volume explores the subtleties of these various issues: Joan Stone discusses language and math issues from the perspective of a researcher who studies mathematics learning among deaf students. In such an approach, language-symbol system issues can be studied for similarities and differences with language-variation issues among both hearing and nonhearing learners of mathematics.

This volume includes discussions of cognitive-developmental changes in mathematics learning; influences of learning contexts, such as the opportunities students have to learn math and exposure to the material or the availability of learning aids and resource materials; teaching contexts, such as the expertise of teachers who are trying to deliver information about math subject matter and make it interesting to the students; the role of language in both teaching and learning, including effects of home language on learning. Discussions are often aimed at pinpointing special learning issues associated with language minorities and with competing language systems. The reader will see that research models and research designs are also points of confluence and divergence in writers' interpretations of research findings. Secondary analyses that are reported emphasize the effects of primary language, home language, and language proficiency. Differences in opportunities to take advanced math courses and encouragement from background cultures to take courses in certain subject matter areas are the focal points for some authors in this volume. In summary, variety in emphasis

TABLE 1.1
Summary Characteristics of Relevant Databases

Database	Socioeconomic Status		Child Characteristics			Non-English Language Background		Language Proficiency	
	Level	Major Variables	Race	Grade Level Versus Age	Sex	Child Language	Home Language	English	Naive
(1) California Assessment	Individual	Five point SES scale filled out by student's teacher.	No	Grade level and age.	M/F	Reported by teacher			
(2) ESEA Title VII Spanish/English Bilingual Education Impact Study	Individual	Teacher completed a form stating parents' education and occupation		Grade and age.	M/F	?	?	English proficiency scores for some students	
(3) Florida Assessment	School	Free lunch, AFDC, N&D, F	Teacher reported up to 3rd grade. Higher levels are self-reported.	Grade level.	M/F	Yes			
(4) High School and Beyond	Individual	Mother's and father's education and occupation, family income.	Yes	Age and grade level.	M/F	1. First spoken 2. Usually spoken 3. Spoken to friends (+).	Language spoken at home (+).	Self-reported for reading, speaking, writing, and understanding.	Self-reported for reading, speaking, writing and understanding.
(5) IEA Math (1982)	Individual	Parents' occupation and education	Yes	Age and grade level.	M/F	Self-reported. Questions asked of student concerning differences between home and school languages.			
(6) IEA Math Study (1964, International)	Individual	Father's and mother's educational attainment and occupation.	No	Grade level and age.	M/F	Assess from county of residence.		No	No

(7) Instructional Accomplishment Information Systems		None	Yes	Grade level and age.	M/F	Self-report on what child's main language is.	Yes	Yes
(8) Migrant Data Beverly McConnell	Individual	Family income for some. Parents were all employed as migrant workers. Parents' education.	Not on test file. Almost all Mexican or Mexican-American.	Age and grade level.	M/F	Primary language		Vocabulary test.
(9) National Assessment of Educational Performance	Individual	Mother's and father's educational attainment, newspapers received regularly, more than 25 books at home, two or more cars or trucks that run, etc.	Yes	Age and grade.	M/F		No	English/Non-English for 17 year olds only (may be able to link with Hispanic question)	No
(10) New Mexico Public Schools		None	Yes	Grade level and age.	M/F	No	Yes	No
(11) New York City Board of Education Data		None	Group level	Grade level and age.	M/F		Some	Data are available on which level of a test a child has taken.
(12) New York State: John Stiegelmeyer	Group, by district. Scored locally	Unknown	Unknown	Grade level.				
(13) NLS	Individual	Mother's and father's education; parents' income; father's occupation; household items.	Yes	Age and grade level.	Yes		Language most spoken in home.	4 point scale on proficiency. Number of hours received of remedial language training.

(continued)

TABLE 1.1 (Continued)

Database	Socioeconomic Status		Child Characteristics			Non-English Language Background		Language Proficiency	
	Level	Major Variables	Race	Grade Level Versus Age	Sex	Child Language	Home Language	English	Native
(14) Project Talent	Individual	Father's occupation, mother's and father's educational attainment, family income, house price.		Grade level and age.	M/F	No	What language regularly spoken at home?	No	No
(15) Relative Language Proficiency Types: A Comparison of Prevalence, Achievement Level and Socioeconomic Status (Incomplete)	Individual								
(16) San Diego Unified School District	Group	Classified by school: level of income of families. Also classified at district level				Yes	?	Letter sent home. 15-minute assessment instrument. Classified as LEP or FEP.	
(17) Sustaining Effects	Individual	Mother's and father's educational attainment and occupation, family income.	Yes	Age and grade level.	M/F	Estimate by language used by parent to help with homework.	1. English spoken at home. 2. Other language spoken at home. 3. Teacher estimate of non-English home language.	No	No
(18) Texas Assessment of Basic Skills	Individual	Free lunch prgram.	Yes	?	M/F	Percent enrolled in Spanish-English classes		No	No

	Achievement			General Academic Performance		Test Language		Teacher Characteristics		
	Math Achievement									
	Computational	Word Problems	Other or Unknown	Reading	Other Indicators	English	Native	Language Proficiency	Certification/Degree	Ethnicity/Race
(1)			Unknown	Yes		Yes		No	No	No
(2)	CTBS Test in mathematics			?	Whichever language most appropriate.		Unknown		Yes	
(3)	Basic skills test written to fulfill Florida minimum competency standards.			Yes	Writing	Yes		No	No	Yes
(4)			Standardized (prepared by ETS)	Yes	Self report of school performance	?	?			Yes
(5)	Both computational and word problems, using 2nd IEA Math Test developed in cooperation with other countries. Achievement measures are based on what teachers taught.					Yes			Yes	
(6)			Specially developed IEA math achievement test.			Yes			Yes	
(7)			Specific tests developed for each district.	Yes		Whichever language most appropriate.				

(continued)

TABLE 1.1 (*Continued*)

	Achievement			General Academic Performance		Test Language		Teacher Characteristics		
	Computational	Math Achievement Word Problems	Other or Unknown	Reading	Other Indicators	English	Native	Language Proficiency	Certification/Degree	Ethnicity/Race
(8)	"Wide Range Achievement Test"			Yes		Unknown		Yes. Language spoken most.		
(9)	Yes	Yes		No	Science	?		No	No	No
(10)			CTBS	Yes		Yes				
(11)			Stanford Math test.	Yes		Yes	At high school level test may be taken in alternative languages.	Teacher data are not available on this file.		
(12)	New York State Pupil Evaluation Program					Yes		?		
(13)			Unknown form of standardized test.						No data	
(14)			Standardized	Yes					Yes	
(15)										
(16)	Children tested only on those levels they have mastered.					Usually	For those in Spanish classes.			
(17)			CTBS		No	Yes		No		
(18)			Texas Assessment of Basic Skills Test.		Vocabulary	Yes			Yes	

	School and Classroom Characteristics				Sample Characteristics		
	Availability of Native Language Learning Materials	Per Pupil Expenditures	Time Spent in Math Instruction	Racial/Ethnic Composition	Sample Size	When Collected	Target Population
(1)	Some data may be available, but actual measures are not known.		Available for school		30,000	1980	Grades 3, 6, 12.
(2)	Unknown	Yes. For both Title VII and non-Title VII classrooms.	By whole class, group and individualized instruction as percent of total mathematics instruction time.			1975–76	2nd–6th grade students.
(3)	No	Data available in Profile Book.	No. Statewide minimum standard.	School level	100,000 for each grade.		Grades 3, 5, 8, 11, Part I. Grade 1, Part II
(4)	No	Yes	Number of courses taken.		58,270	1980	All high school seniors and sophomores.
(5)	No	No	Yes		8th–9,000 12th–8,000 (only those with four years of high school math).	1982 (available late 1982)	Grades 8 and 12.
(6)		Yes			?	1964	Data from 12 countries.
(7)	In some of the samples.	Yes (at district level).	Yes		SWRL: some district data		Elementary school children in selected districts.
(8)	Some materials available to all children.	Unknown	No	Mexican, Mexican-American, Anglo.	1,020	1980	All children specifically enrolled in McConnell's bilingual program (pre K to 3).
(9)	No	No	No		17,000 (age 9), 26,000 (age 13), 26,000 (age 17).	1977–78	
(10)	No	Yes (not on a data set).	No		All students.	Yearly	Grades 5, 8, and 11.

(continued)

TABLE 1.1 (*Continued*)

	School and Classroom Characteristics				Sample Characteristics		
Availability of Native Language Learning Materials	Per Pupil Expenditures	Time Spent in Math Instruction	Racial/Ethnic Composition	Sample Size	When Collected	Target Population
(11) No	No	No	No	All students in NY city schools.	Collected regularly.	All students in NY city schools.
(12) No		No				100% of all 3rd and 6th graders (state has only district level data).
(13) No	No	Number and type of courses provided.			1972	High school seniors.
(14) (15) No	Yes	No		440,000	1959–60	Nation's high school youth.
(16) Available in Spanish, Portugese, and Vietnamese.	No. At district level only.	In accordance with the curriculum guidelines for entire district.	Majority of LEP students are Spanish and Indo-Chinese.	10,000 (elem) 5,000 (sec)	1975–76	LEP elementary and secondary. AGP group.
(17) No	Yes	Yes (individual level)	Yes	15,579	1975–76	All children enrolled in grades 1–6.
(18) Yes (K–3)	No	No		All students.	Yearly	Grades 3, 5, and 9.

Note: This table was prepared by D.E. Myers and A.M. Milne under contract 300-80-0778 from the U.S. Department of Education, under Part C: Math and Language Minority Students Project.

and variety in perspective characterize what follows. What the authors unanimously state is the need to study the issues together and to infuse them with our collective expertise.

OVERVIEW OF THIS VOLUME

This volume is organized into three sections, each comprised of five chapters. The focus of the first section is on the linkages among mathematics, language, and learning. The chapter by Cocking and Chipman lays out the range of issues that need to be addressed in assessing the state of knowledge of the role of language in mathematics learning. Saxe raises the importance of developmental issues by asking how culture imparts basic knowledge of number concepts that the child enters school already knowing. Scholnick builds upon the developmental theme and presents a research agenda for developmental psychologists interested in the study of thinking. She points to the broad range of research questions that need to be addressed by those who maintain that the acquisition of any form of literacy may foster a common set of skills or competencies: learning, facilitation, interference, and variations in the complexities of academic content. Stone presents a discussion of symbol-system conventions in learning language and mathematics and discusses the interface of the two with respect to deaf learners.

The second section of the book consists of discussions of cultural and linguistic influences upon mathematics learning. De Avila suggests that school achievement of language minority students can be understood as a function of three interacting factors: a) intelligence and cognitive style, b) interest and motivation, and c) opportunity and access. Leap's analysis of native Americans (Ute) illustrates how students evolve problem-solving strategies that derive from assumptions imparted by the culture in which they live. Charbonneau and John-Steiner pick up the themes of culture and influence of daily life activities as the underpinnings of performance. They discuss the role of patterns that stem from the environment that influence higher order cognitive skills, such as recursiveness and estimation that derive from patterns and geometric relationships. Tsang's analysis explores assumptions that Asian Americans exhibit few problems in mathematics learning and considers how language barriers are overcome in this group of learners. Tsang points out that the currently popular variables, time-on-task and awareness-of-employment-opportunities, are sensitive to cultural and ethnic differences. MacCorquodale's chapter goes beyond research on gender and achievement studies to examine gender and ethnicity simultaneously in order to determine the additive effects on students who have double disadvantages in their learning opportunities.

The third and final section of this volume contains discussions of research models and data analyses. The chapter by Spanos, Rhodes, Dale, and Crandall builds upon research that demonstrates high correlations between math achievement and English reading ability, and reading comprehension skills and performance on deductive reasoning tasks. These researchers study the influence of language register, which contrasts special content-specific terms of language that are commonly used in the classroom with the special terminology of mathematics. Mestre also looks at the issues of interference between the mathematics register, as defined by Spanos et al., and common written language expression and how this confusion leads to errors in problem-solving by Hispanic college students. Like Leap's analysis of Ute Indian students, Mestre's chapter is an analysis of problem-solving strategies used by Hispanic students as they try to work through math and reasoning problems. Duran builds upon the issues confronting college-age learners and upon the strong relationship between reading comprehension skills of Puerto Rican students and their performance on deductive reasoning tasks in English and Spanish. The chapter by Myers and Milne returns to issues of home language and primary language of students, as well as language proficiency, as they present a broadly scoped study and research model for secondary analyses. Finally, Travers surveys some of the findings of the Second International Mathematics Study and outlines some of the kinds of analyses that help explain achievement patterns in junior high school mathematics learning. Specifically, Travers contrasts *intended* curriculum with *implemented* curriculum in his analysis of achievement differences.

ACKNOWLEDGMENTS

Many people have contributed to this volume whose names to not appear in the Table of Contents. Some of the people who clarified our thinking as we interacted at conferences and who read drafts of these chapters are: David Berliner, Robert Boruch, Edward Esty, Jack Lochhead, Amado Padilla, and Blanca Rosa Rodriguez. Lois-ellin Datta initiated this project in her capacity as Associate Director for Learning and Development when she was at the National Institute of Education, and she facilitated broadening the scope of the project by involving various parties, namely NIE, the Office of Bilingual Education and Minority Language Affairs, and the Department of Education's Office of Planning, Budget and Evaluation. People from these latter two offices included Gilbert Garcia (OBEMLA) and Alan Ginsburg (OPBE) who jointly funded the project with NIE. Shirley Jackson, during her tenure as Associate Director for Learning and Development at NIE, lent her assistance by seeing that this project continued through completion. Frank Murray, as Dean of the College of Education at the University of Delaware, provided many support services to the first author, which helped move this volume along. Similar support was provided to the second author by Frederick Byron, Dean of Natural Sciences and Mathematics at the University of Massachusetts. While we certainly appreciate the assistance of these people, the views expressed in this volume are those of the individual authors and do not reflect the opinions of the various people or institutions who supported the project.

Chapter 2

Conceptual Issues Related to Mathematics Achievement of Language Minority Children

Rodney R. Cocking
National Institute of Mental Health
Susan Chipman
Office of Naval Research

LANGUAGE AND POVERTY

Almost 2 decades ago, a monograph came out of the Institute for Research on Poverty that dealt with some of the practical questions about the language of children in antipoverty programs (Williams, 1970). The gap that often separates research from practice was narrowed when educators, linguists, and social scientists pooled their expertise to grapple with the relationship of language behavior and school learning. The conclusions that were drawn from the collection of papers in that monograph remain vital concerns today:

- We must avoid confusing language differences with deficiencies;
- It is a reasonable and desirable goal that all children in the United States be able to function in Standard English in addition to whatever language or dialect they have learned in the home;
- We must develop new strategies for language instruction, as our existing ones are largely inadequate for use with children coming from varying language backgrounds;
- We must increase our research efforts in the study of language differences in the United States, and the interrelation of these with different social and family structures;
- Language programs for the poor should incorporate research and evaluation components.

In the course of this chapter, we consider the extent to which the alleged impact of language minority status upon school achievement is a function of language per se, or of other factors, such as poverty. Certainly, there have been reports that have chosen to regard achievement deficits as determined by single clusters of interrelated variables, such as language variables (Williams, 1970), or variables associated with low SES (Bruner, 1975; Bruner & Kenney, 1965). Bruner's conclusion was that child development and later intellectual competence (e.g., problem solving, goal seeking, use of language, etc.) are adversely affected by poverty, although he did not discuss specifically the case of language minority children. It may be due to a focus upon single-factor or single-cluster models that two decades of research have not brought us much closer to understanding the relationship between language and school achievement.

Wider perspectives on the issue of the relationship between language and cognition should be considered, in contrast to correlations between language and single skills, such as reading. Ethnicity may be an important variable if the relationship between language and achievement cannot be treated in general terms but must be qualified with respect to the circumstances of particular language groups. Teaching and instructional variables are also significant, and as we gain understanding of teaching influences, we can only expect their significance to become greater, since responses to individual differences are likely to lead to more complex schooling practices.

Developmental psychology also provides broader perspectives on educational issues. Piaget's and Bruner's theories teach us both that learning and understanding are different at various points in development, and that instruction given at one stage should prepare the child for what is to follow. As children's understanding increases, their capacity for more advanced learning also increases. Educationally, this means that with growth and proper preparation, a curricular area such as mathematics becomes more differentiated over the school years. Not only do we learn more math as we progress through school, but the mathematics we learn is of a different type at eighth grade as compared with third-grade math. The student's expanded mathematical knowledge, as well as growth on a variety of psychological and cognitive dimensions, must figure in to our analysis of the role of language in learning and what is learned. What we learn about mathematical growth, both in terms of amount of knowledge and kind of information, is likely to tell us something about the optimal timing and methods of instruction.

Although it is true that the issue of language and cognition is larger than any specific curriculum topic, the present chapter begins by taking a look at the major issues of mathematics achievement with respect to language minority students. The purpose is to look at the relationship between mathematics learning and language skills. We then ask if our conclusions are unique to mathematics and if the effects are due to language backgrounds of the learners.

SOME ASSUMPTIONS AND QUESTIONS

Some general assumptions are made about math achievement and bilingualism that need to be reconsidered in light of the evidence that has come out of research and program evaluations. These assumptions can be converted into questions that need to be addressed in studying the issues. As we review these questions we have to ask repeatedly what we know, the adequacy of our knowledge for today's problems in education, and whether the current state of knowledge challenges the assumptions that are widely accepted about the influences of language upon math learning. The following questions need to be addressed:

> What is known about the mathematics achievement among language minority groups?
> What is known about the effects of language upon mathematics achievement?
> What is known about the dependence of mathematics learning upon language skills?
> What are considered to be the important moderator variables in the math/language relationship, and what is known about each?
> What happens to the math/language relationship in the course of development?
> What is known about the determinants of mathematics achievement in the general population, where language minority status is not an issue?
> What do we know about instruction, both in terms of the quality and specifics of proposed remedies and in terms of impact upon academic achievement?
> How general is the problem of math achievement deficits in language minorities, and does the problem seem to have different patterns of relationships in various ethnic groups?

The assumptions about math learning and language influences suggest a particular model of how learning is influenced by both linguistic and nonlinguistic factors. Such a model can be displayed in the following way to address the three major categories of influences upon school learning: Entry characteristics of the learner; Opportunities provided to the learner; and Motivation to learn.

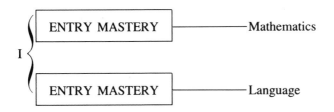

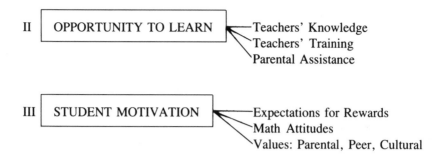

II | OPPORTUNITY TO LEARN | ←——Teachers' Knowledge
Teachers' Training
Parental Assistance

III | STUDENT MOTIVATION | ←——Expectations for Rewards
Math Attitudes
Values: Parental, Peer, Cultural

This model provides a way of conceptualizing the relevant variables in the language-mathematics relationship in a way that takes account of both Developmental Status and Linguistic Capacities for receiving and utilizing classroom instruction. The model can be expanded along lines of Input to the children and Output, or Mastery, which is child performance. The schematization in terms of Input and Output follows in the diagram presented here.

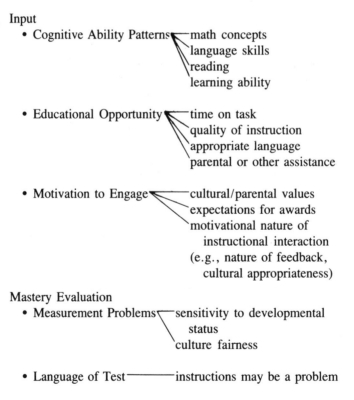

Input
  • Cognitive Ability Patterns ←——math concepts
                                 language skills
                                 reading
                                 learning ability

  • Educational Opportunity ←——time on task
                               quality of instruction
                               appropriate language
                               parental or other assistance

  • Motivation to Engage ←——cultural/parental values
                             expectations for awards
                             motivational nature of
                               instructional interaction
                               (e.g., nature of feedback,
                               cultural appropriateness)

Mastery Evaluation
  • Measurement Problems ←——sensitivity to developmental
                               status
                             culture fairness

  • Language of Test ————instructions may be a problem

- Performance Variation ⟍⟍⟍ types of math problems-word versus
  computation problems

  math versus other cognitive skills
  performance

The Input portion of the diagram enumerates the kinds of variables implicated in the model: Entry characteristics of student in the content areas, and the opportunities for learning provided through Teacher skills and Parental support, are represented in I and II of the model. Student Performance (Output) is also influenced by Initial, background learner skills (I) and Opportunities to learn (II). Similarly, Student Motivation (III) is a global variable that figures into both Input (learning) and Output (performance) components of the model. Thus, the model represents classes of relevant variables and the Input-Output representation illustrates how the variables are manifest differently, depending upon whether the issue is one of learning or one of performance. Fuller discussions of the modeling of diverse influences upon mathematics enrollment and achievement can be found in Chipman and Thomas (in press) and Chipman, Brush, and Wilson (1985).

PREVIOUS EXPLANATIONS

In various isolated studies, researchers have offered a variety of interpretations or explanations for the low mathematics achievement found among language minority students. The explanations generally have been of the following types:

- Poverty: Children of poverty, generally, do not achieve at the levels of nonpoverty children. These explanations assert that low socioeconomic status might contribute to the achievement problems of language minority students.
- Language: Perhaps children do not understand their school instruction. In the case of math, they may not understand the language of the instruction and/or word problems. These explanations raise questions about the role of language proficiency and the language of instruction and whether or not math learning is independent of language skills.
- Culture: Values, parental assistance, and motivation are offered as explanations for the lower performance levels of language minority students.
- Cognitive Abilities Patterns: Perhaps learning tempos and cognitive styles are different for various ethnic subcultures. Explanations of this sort suggest that matching of teaching style to learning style is needed.

These explanations relate to the model previously presented. Poverty, language skills and cognitive abilities are input or entry characteristics. The cultural milieu is an aspect of the Opportunity to Learn either at the home-culture level of support or the institutional level, where ghetto schools may have different student/teacher ratios, perhaps teachers of lesser training, or even teachers unable to speak the student's home language. The goal is to derive a research model that incorporates all these classes of variables taken together rather than as clusters guided by the assumptions of poverty, language, culture, or cognitive ability patterns.

QUESTIONS FOR REVIEW

There is a wide range of questions that need to be raised and addressed before accepting the assumptions made about the learning abilities and achievement patterns of language minority students. The purpose of the following section is to look at each of the questions separately.

What Determines Math Achievement in the General Population?

In the general population, much research has been done investigating the relation of math performance to general intellectual ability or to more specific abilities. An early review of work through 1960 (Feierabend, 1960) looked at 19 studies concerning the relationships of general intelligence and special abilities for mathematics. It was concluded in that review that high general intellectual ability is important in math, but that questions about the existence of specific abilities could not be resolved conclusively with the available research.

Similar attention has been given to possible relationships between mathematics performance and specific verbal or language abilities. In such studies, reading test scores typically have been used as the indicator of verbal ability. Reading ability positively correlates with scores on math problem solving, but the total picture of abilities must be considered. The positive correlations between reading and math that appear to hold throughout the grade-school age range drop markedly when their common overlap with general intelligence is partialed out. Thus, the positive relationship between reading and math can be explained largely by the overlap of these two variables with general intelligence (Aiken, 1971a).

Based on factor analyses, Wrigley (1958) concluded that, among college students, high general intelligence is the first requirement for learning mathematics successfully. Wrigley also found orthogonal math and verbal group factors that led him to conclude that there are distinct verbal and math abilities. The positive correlation between verbal and math performance could be attributed to their common relationship with general intelligence. The conclusion that Wrigley drew was that "the portion of verbal ability which is not included in general intelligence is of no importance to mathematics achievement" (Aiken, 1971a, p. 304).

Although general intelligence can account for a proportion of the common variance between math and verbal ability, Aiken (1971b) points out that a large portion of the variance is left unexplained. Using the same studies that Wrigley reported, Aiken compared the zero-order correlations with corresponding first-order partial coefficients. The comparison continued to support Wrigley's hypothesis, but Aiken points out that there is more to the story: the majority of the first-order partial coefficients are significantly greater than zero. Additionally, the correlations between math ability and general intelligence drop considerably when reading ability is partialed out.

Clearly, the dimensions of math ability need to be studied further. What is of importance here is reaching some conclusions about the interactions of math and verbal abilities. Where correlations indicate significant relationships, the next step is to probe why the variables are related. Obviously, the significance and appropriate interpretation of standard verbal ability and reading measures may be quite different in populations whose language experience is divided among two or more languages than it is in the typical monolingual population (Duran, 1985). Ordinarily, for example, a measure of vocabulary functions effectively as a measure of general intellectual ability; it would not do so in a population that mixes native and non-native speakers of the language tested. More will be said about the specific relationship of reading and verbal abilities with math ability in another section.

As yet, research provides no reason to assume that inborn differences in general ability account for minority-majority group differences in achievement. Specifically in the area of mathematical concepts, Ginsburg (1981) has provided evidence that the most fundamental mathematical concepts are equally present in children from disadvantaged backgrounds. However, even at the preschool level, there are differences in communicative competence that presage later measured differences in school achievement. More attention needs to be given to the organismic variables—such as development—and to the environmental variables—such as home and school opportunities for learning—that determine what kind of mathematical competence is ultimately built upon these shared conceptual foundations.

What is Known About Mathematics Achievement in Language Minority Students?

Two types of evidence bear on the question of math achievement among children of language minority status. There are cross-sectional data that tell us something about these learners relative to majority students at various grade levels. There are also data that tell us how the problem has or has not changed over the past few years.

Data from the National Center for Educational Statistics in 1978 (Highschool & Beyond) indicated that Hispanic youths were significantly below majority

TABLE 2.1

AGE	9	13	17
Math skills			
Knowledge & Skills	−7.9%	−12.0%	−12.0%
Understanding	−8.2%	−11.8%	−13.8%
Application	−8.2%	−10.5%	−12.1%

students on three areas of math achievement including knowledge and skills, math understanding, and application. The data also indicated that the problem was more pronounced among older than younger students, as Table 2.1 shows. From the table, it appears that there is a growing deficit as the children get older. The same finding occurs for native American students, and as they move to higher grades, the gap is greater than for any other ethnic group.

Although a variety of languages are involved, native Americans are treated as a language minority group in evaluating educational progress. They often come to school with limited skills in English and at home many speak and/or hear a predominantly different language from English.

Zimiles (1976) found that the achievement gap appears early in the education of native Americans, a deficit that has severe consequences by the time the students reach high school. Zimiles reported that 57% of the second graders in his sample were below the 25th percentile on math, and that by third grade the deficit grew to 81% falling below the 25th percentile. Fourth and fifth graders were performing not quite two grade levels below the norm, but high schoolers were more than two grade levels below the norm. The differences were greatest between Grades 10 and 12.

Developmental differences aside, have there been any changes in math achievement patterns of minority students in the past few years? The National Assessment of Educational Progress (NAEP) report (1979) indicates that achievement remained constant between 1973 and 1978 for Hispanic 9-year-olds. Thirteen-year-olds, by comparison, showed a decline between 1973 and 1978, with a difference between whites and Hispanics reaching its maximum of 15 percentage points.

The same math achievement differences that occur for language minority students during the elementary and secondary school years are apparent when they take college and graduate school examinations. Mean SAT scores for Hispanic students are 78 points below white students (412 vs. 490). Quantitative and analytical scores on the Graduate Record Examination (GRE) show an even wider gap:

TABLE 2.2

Graduate Record Examination	Hispanic	Anglo
Quantitative	438	538
Analytical	427	538

There is little doubt about the existence of math achievement differences that appear at least by the mid-elementary grades and continue to increase into high school and beyond. These differences appear to occur earlier in some ethnic groups than others, and the differences appear to be greater for some groups than for others. To answer the original question of "What is known about mathematics achievement in language minority students," we have to conclude that although there is no evidence to suggest that the basic abilities of minorities are different from Anglo-speaking students, the achievement differences between minority and majority students are pronounced. Ginsburg (1981) indicated that the vocabulary children have for expressing math and number concepts differs widely. We are led to ask, then, Is language a factor in these achievement differences?

What is Known About the Effects of Language on Mathematics Achievement?

It has been asserted that the math achievement deficits documented for native Americans and Hispanics occur in the course of schooling because these students encounter special learning difficulties associated with differences between the language used in school and their dominant or home language (Merino, 1983). This problem bears looking at from two angles: first, is the evidence concerning the relationship between language and math achievement. Secondly, there is the more causal question of what is known about the dependence of math performance upon language skills.

In the general population, a positive relationship between math achievement and verbal ability has been documented over the school grade range (Aiken, 1971a, 1971b; Cottrell, 1968; Goodrun, 1978). The strength of the association seems to decline in the course of schooling, but remains significant throughout, as indicated in Table 2.3.

What do we know about the relationship between math learning and the minority student's language skills? Some groundwork in recent years offers preliminary insights, if not answers, into the problem. The results, however, are by no means conclusive or even consistent, in many cases. De Avila (1980), for example, concluded that language proficiency is not strongly predictive of math

TABLE 2.3

Grade Level	r	Source
3	.86	Cottrell, 1968
4–6	.41–.40	Cleland and Toussaint, 1962
6	.60	Erickson, 1958
11	.53	Pitts, 1952

achievement in Grades 1, 3, and 5 for Hispanics. Mestre (1981), in contrast, reported positive and significant correlations between problem solving and language proficiency among Hispanic bilingual college engineering students. This comparison illustrates the number of dimensions on which studies differ when ostensibly similar populations have been studied. The conflicting results may reflect age and developmental differences that exist when comparing studies within a given ethnic group. The results can also be attributed to selection of subjects in the studies. It is not clear whether Mestre's results would hold beyond the subgroup of engineering students. Engineering students tend to have higher math than verbal scores on the Scholastic Aptitude Test, whereas in the general population the two sets of scores are similar. It is also possible that a different correlational structure would emerge within the college student sample if monolingual students had comparable SAT scores to those of the bilingual students studied. The first effort needs to go into looking at data from various points in the schooling process to assess developmental trends.

One data source is the National Assessment of Educational Progress (NAEP). Mathematics data included both computational and word problems, but language skills and language achievement data were limited. For example, one does not know about the students' skills (proficiency) in either English or native languages. Reading data, which might index language achievement, were also lacking. The language in which the test was administered (English or native) was not reported. The sample included 9-year-olds, 13-year-olds, and 17-year-olds, and the samples were sufficiently large (17,000 9-year-olds; 26,000 13-year-olds; and 26,000 17-year-olds) to produce the following trends:

- At age 9, reading and mathematics scores of Hispanics show about the same deviation from the national norms (1980); in the past (1970/1975), mathematics was slightly better than reading, due to lower reading scores in the 1975 assessment than in 1980.
- At age 13, however, although the reading scores remain at about the same level (1970/1975) or decrease somewhat relative to 9-year-olds (1980), the mathematics scores for both assessments decline, creating a reading-mathematics disparity with lower scores in mathematics.
- By age 17, the gap between reading and mathematics is 3 percentage points

difference, with lower mathematics than reading scores for both 1970/1975 and 1980.

Conclusions from the NAEP (1980) data were that

- Reading has improved between 1975 and 1980 for Hispanic students, although performance is still considerably below the national norms;
- Mathematics has not improved for the 9- and 13-year-olds and the improvement for the 17-year-olds, although encouraging, continues to reflect a reading/mathematics gap.

These data, then, suggest that there may be some special problems in mathematics achievement of Hispanic youth, particularly at the upper age levels.

Other data reported for sixth graders in California add to the complexity of the picture. The data are consistent with the NAEP report at the 9-year-old level for Hispanics (i.e., low scores in both reading and mathematics relative to national norms, but no math/reading differential). However, other language groups (non-English and non-Spanish) showed a math performance score that was higher than the reading scores.

The California testing was conducted in English and a measure of English reading achievement was used to indicate language skills attainment. The California data base did not distinguish math performance that was language-based (word problems) from nonlanguage-related mathematics problems (computation problems) and oral language proficiency was not determined for the child's knowledge of the native language or for English. The group differences, therefore, could be attributable to either specific language interference issues (Spanish), mathematics learning opportunities, or factors of language proficiency. Because word problems were not differentiated in the math assessments and because English proficiency was not determined, the study shows only that reading achievement is related to test performance among non-Hispanic speakers. Thus, language proficiency (both native and English), opportunity to learn, and specific symbol-system interferences between Spanish and mathematics remain unknowns in the language-mathematics equation. However, the California results replicate the NAEP finding of below-norm performance and comparability of math and reading skills levels for 9-year-old Hispanic students.

Data from the Sustaining Effects Study (SES) also indicate that at the lower grades there is not a mathematics/reading disparity for Hispanic youth. The Sustaining Effects Study data focused on multiple predictors of child language proficiency (whether English is spoken at home; what language other than English is spoken at home, and a teacher estimate of non-English home language). The report also distinguished home from school influences (extent to which parents helped with schoolwork), and looked at achievement test data as well as reading achievement (CTBS). Children in the Sustaining Effects Study were evaluated in English only and no actual language proficiency data were collected for native or English languages. The investigators assert, however, that the variety of language indicators (such as home vs. school language) led to finer discrimination of language-related performance, as well as subtle developmental changes that could be detected by looking at all children (15,500) in Grades 1 through 6, rather than by means of cross-sectional sampling procedures used in other data bases. But, once again, the contributing variables were too complex to enable one to conclude that the developmental trend is tied in with schooling processes alone. For example, home-based parental assistance is likely to cease once students advance to topics out of the parents' own school experiences or abilities, contributing, in part, to the decline in achievement at the upper grades.

Is all the evidence negative? Both Lambert (1972) and DeAvila (1980) report positive benefits of bilingual status on cognitive development. Lambert attributes his results to a measurement design feature: he used nonverbal intelligence tests to measure the effects of bilingualism on IQ. DeAvila reported two findings of importance: first, as Hispanics' oral skills in Spanish improved, their math achievement scores began to decline. This led him to propose that effort expended on language mastery may interfere with mathematics learning. Secondly, DeAvila found that if one looked at levels of proficiency within mono- and bilingual groups, the most proficient bilinguals outperformed both less proficient bilinguals and monolinguals on cognitive styles measures such as the cognitive components of Witkin's Field Dependence/Field Independence. This study is notable for the relative adequacy of its treatment of language proficiency, but the DeAvila result regarding the progressively negative relationship between oral language skills and math achievement calls for special attention. To probe this result, we turn to a second aspect of the math/language relationship and ask to what extent math performance is dependent upon language skills.

What is Known About the Dependence of Mathematics upon
Language Skills?

It is one thing to find evidence for a relationship between two cognitive skills, such as math and language, but it is quite another matter to be able to show that one of them is influencing the other. Although the evidence must be built from inferences, we can look at the problem from what we know about how language influences performance on math tasks, and then take a closer look at moving between math and English from the perspective of a linguistic model. This latter tack will help us understand math as a language and thereby provide further evidence needed for an analysis of the effects of language skills on math performance.

Cossio (1977) looked at the issue of the dependence of word problem solving upon college students' language development and found that the fully bilingual students with a good math background performed better on the English version of the Mathematics Placement Test than they did on the Spanish version. Perhaps this was because they had been instructed in English. In contrast, for those who were not fully bilingual (i.e., more Spanish dominant), the Spanish version produced the higher scores. Macnamara had previously reported lower performance for bilinguals when the language of instruction was the weaker of the bilinguals' two languages (Macnamara, 1966).

Verbal ability may be the issue and not bilingualism, per se. There are students who have high verbal ability and still perform poorly on math tasks. Nevertheless, those students who have low verbal abilities more often than not also show poor math achievement. This is not surprising because math problems have to be read. The question is whether or not reading about mathematics presents any special problems. Spencer and Russell (1960) claimed that the difficulties in reading mathematics are due to the specialized language for expressing ratios, fractions, and decimals; that the names and terms are unique to mathematics; and that reading about computation requires some specialized procedures. More recently, several researchers have demonstrated just how pervasive the problem is for translating certain math problems from English into mathematical representations. They point out that students first have to convert the English statements into math statements before attempting to solve the problems. For bilingual students, this process may involve a third layer of translation.

Studies by Rosnick and Clement (1980); Clement, Lochhead, and Monk (1981); Kaput and Clement (1979); and Rosnick (1981) all report the widespread inability of university engineering students to translate relationships expressed in natural language into corresponding mathematical expressions, and vice versa. For example, 37% of engineering students could not write in mathematical language the following statement when it was presented in English: There are six times as many students as professors. To underscore the pervasiveness of the problem of not being able to translate forth and back in mathematical and natural languages, 57% of the nonscience majors could not make the translation. The most common error was to translate the mathematical symbol $[=]$ to read "for every" rather than "is numerically equal to." Thus, the expression was usually rendered $[6S = P]$ rather than as $[S = 6P]$. Seventy-three percent of the students in the Rosnick and Clement (1980) study made the same translation error when representing the following statement in mathematical terms: At a certain restaurant, for every four cheese cakes that were sold, five strudels were sold. Again, the most common mistake was to reverse the variables and to write the relationship as $[4C = 5S]$.

Lochhead (1980) believes that such translation difficulties are not restricted to high school and college students. He found that high school teachers and college faculty also are poor at representing math problems. About 46% of the high school teachers and 40% of the university faculty erred on these problems. Even physics, chemistry, math, and engineering professors had an error rate of 14%, whereas high school physical science teachers had an error rate of 39%.

Clement, Lochhead, and Monk (1981) interviewed 15 students in an effort to understand the translation issue better. The subjects made the kinds of translation errors mentioned previously, but the researchers ruled out sheer carelessness as the cause. Rather, misconceptions about how words map onto the math symbols, difficulties with word-order matching, and so forth, appeared to be at the heart of the matter. The authors conclude that students have not learned how to conceptualize within the language of mathematics (Lochhead & Mestre, in press). For example, "There are six times as many students as professors" becomes $6S = P$ by a direct sequential word-for-symbol mapping of words onto symbols. The translation errors have been replicated in students from Fiji, Israel, and Japan (Mestre & Lochhead, 1983) and thus are not unique to Anglo and Hispanic students in American schools. A similar transformational rule error also has been seen in children's early attempts to map action sequences onto syntactic transformations, such as the passive voice (Cocking & Potts, 1977; Potts, Carlson, Cocking, & Copple, 1979).

Morales, Shute, and Pellegrino (1985) relate mathematics to larger conceptual issues of representing word problems. The authors concluded that language, as a representational vehicle, did not affect performance: they found no performance differences between English and Spanish versions of a word problem test for Hispanic students. Errors that occurred were conceptual, not calculational. The issue, as Morales et al. see it, is one of accessibility—whether performance is in the dominant language. When it is, they say, performance transcends the language differences. What this means is that language and representation become relevant factors when the language of performance is not the dominant language.

Studies repeatedly have shown that algebraic word problems are solved with greater success when instruction focuses on meanings of the words and their translation into corresponding mathematical symbols (e.g., Call & Wiggen, 1966; Gilmary, 1967; Madden, 1966; Rose & Rose, 1961). Although other variables are always operating in training studies, the results of these studies point to strong effects of language upon mathematics achievement, an effect so strong that some posit a dependence of math achievement upon proficient language skills of a special kind. Both ordinary language and the language of mathematics are being used to represent the same deep semantic structure of relationships among quantities. Mastery of those semantic structures, however, could be considered a distinct conceptual or intellectual achievement. We return to the issue of language and representation again when we take up the topic of Opportunities to Learn.

What are the Important Intervening Variables when Studying Mathematics Achievement in Language Minority Students?

In middle-class majority language children, the switch between home and school languages results in high levels of functional bilingualism and high academic achievement (Lambert, 1972). Yet, for many minority language children the same transition seems to lead to inadequate command of both first and second languages. Why? We have to ask what is known about other intervening variables that might affect language and, in turn, affect the mathematics achievement of bilinguals. Additionally, the same factors that affect language skills directly or collaterally may directly impact on other areas of academic achievement. In order to explore some of these variables, both sociocultural factors and school program variables need to be addressed. Without considering at this point which are determiners and what is outcome (e.g., sociocultural and program variables as determiners of language; or alternatively, language and sociocultural variables as determiners of program effectiveness, etc.), we can look at what is known about effects of five classes of variables upon verbal and mathematical skills. These five groups of variables include social class, or socioeconomic status, (SES); parental encouragement and assistance; teacher competencies; opportunities for learning in a bilingual environment; and testing for mastery of conceptual learning in minority students.

What is Known About the Effects of Social Class upon Language and Mathematics Skills?

Simmons (1985) pointed out that application of the cross-cultural research model in social-class comparison research has led to some erroneous interpretations of results. For example, social class, or socioeconomic status, is often used to create putatively comparable samples from culturally different groups, and then conclusions that are drawn about group differences are explained as either cultural or class differences. If social class is held constant, the conclusions are in terms of ethnic differences, and so forth. Simmons' argument is that social class indicators are not equivalent across groups, and research studies need to isolate the unique effects of the many variables associated with socioeconomic status measurement before drawing sweeping conclusions about their effects on cognitive performance.

This criticism is directly relevant to the problem at hand because differences in language environments often lead researchers to investigate those environments directly, oftentimes comparing home differences between SES groups. Trotman (cited in Simmons, 1985) points out that the home environments of middle-class children are not all alike: middle-class black home environments are not like the home environments of middle-class whites, and to conclude that we can use a between-group design with socioeconomic status as the independent variable is erroneous. Perhaps this is why variables such as "home environment" or "socioeconomic status" have proved better predictors for some groups than for others. We have to consider whether independent variables have the same relationships for various groups; and, if the relationships appear to hold, to ask further if the strength of the relationships is the same across groups.

Perhaps it is for reasons such as these that socioeconomic status appears to have a weak relationship with school achievement, and why socioeconomic status is more likely to be associated with life goals, self-concept, and so forth (Simmons, 1985). Math achievement is heavily dependent upon school instruction (Wise, 1985), and it is not likely that math achievement would be related strongly to family background variables tied to socioeconomic status. Occupational expectations and information associated with socioeconomic status may affect the value assigned to mathematics study and achievement. The family background variables other than socioeconomic status likely to affect school performance, and especially mathematics, are such things as family member attitudes toward mathematics and mathematicians, and the early experiences the child has in environments that convey these attitudes (Mestre & Robinson, 1983; Morning, Mullins, & Penick, 1980; Ortiz 1983). In the case of Hispanic women and native Americans, Green (1978) and others (e.g., Leap, this volume; Mac-Corquodale, this volume) report that there is a negative image of math, whereby mathematicians are perceived as remote, sloppy, obsessive, and (no pun intended) calculating. No matter what your socioeconomic status happens to be, it is not likely that you will strive hard to become an unscrupulous, calculating, cold, sloppy, and obsessive number cruncher.

In pursuing the social class question, research will have to look with greater detail into the meaning that the components of this construct hold for the groups under study.

What Aspects of Teacher Competencies Affect Math Learning in Bilingual Students?

Teachers' competencies, for this discussion, mean more than their levels of understanding mathematics or language skills. Competencies of teachers will include their abilities for recognizing mathematical talent, encouraging math abilities through instruction, pointing out math-related career opportunities to those who would otherwise avoid them, and so forth.

How effective can teachers be if they are poorly prepared for a discipline or if they don't like the subject matter? Teacher effectiveness as a variable has to be looked at not only in terms of teaching skills, but also for the attitudes teachers convey as role models.

Green (1978) has reported that minority teachers are more likely than majority teachers to state that they dislike math, and it is not unlikely that these negative attitudes are conveyed to the students. Unfortunately, the majority of teachers who like mathematics are not likely to encourage the minority student to take advanced, difficult courses for fear that the student will experience failure. These same teachers extend the protective umbrella in their own math courses by reducing the challenges they set for minority students (Simmons, 1985). Minority students are more likely to receive encouragement to take on challenges such as attending college if they are enrolled in minority schools than are their peer counterparts who are enrolled in majority schools. The problem is, however, that the encouragement is seldom to pursue math (Feierabend, 1960; Simmons, 1985).

In summary, minority teachers may exhibit substandard competencies for encouraging minority students in mathematics education, in addition to presenting role models who themselves dislike math. At the same time, majority teachers do not serve minority students well either because they tend not to set sufficient challenges or to direct the minority student toward advanced math courses.

What observers of second-language classrooms report (cf. Simmons, 1985) is that second-language environments focus primarily on mechanical and decoding skills. Even high-ability students in the second-language classrooms were destined to perform low-level learning exercises on par with the types of learning activities one saw going on among the lower ability learners within the first-language classrooms. As Simmons points out, "this circumstance compounds the difficulties of students who are already deficient in academic skills" (p. 531).

The teacher, as Moll, Estrada, Diaz, and Lopes (1980) observe, is the mediator between curricula, materials, the goals, and the students. Simmons (1985) states: the teacher "regulates the level of difficulty of the lessons by modifying, changing, and adjusting the task demands" (p. 530). Helping children adjust to the problems to be learned and providing the necessary experiences that enable students to generalize their learning in order to be self-reliant learners are particularly difficult teaching demands for instructors working in classrooms that contain bilingual children. The teaching vehicle tends to be language and thus the teacher's communicative competencies are critical (Chan & So, 1982). Research studies on bilingual education program effectiveness seldom contain information about the effect of teacher language proficiencies in either English or non-English. This important gap in the research literature needs to be filled by examining the relationship between teacher language proficiencies and student outcomes. Teacher competencies in content areas such as math need to be examined, as well.

In summary, we cannot overlook the teachers' own language proficiencies when we consider influences on language-dependent subject matter. Merino, Politzer, and Ramierz (1979) found that bilingual teachers' scores on the Teachers' Spanish Proficiency Test correlated positively with their students' scores on English proficiency. The finding that successes in university tutorials for minority students are maximized when the tutors are themselves minority membership staff (Green, 1978; Merino, Politzer, & Ramierz, 1979) suggests that there may be other specific teacher competencies related to student characteristics. What aspects of these interventions account for student improvements? Issues of teacher competencies emerge as being nearly as important as the students' characteristics.

What are the Learning Opportunities for Language-Minority
Students?

Teacher competencies are part and parcel of a larger issue: the opportunities
available to bilingual learners. Low teacher competencies, of course, mean that
one source for opportunities to gain from education is minimized. Administrative
practices can contribute to and, indeed, compound the problem if tenure decisions have been made on seniority rather than on curricular needs of a school.
Green (1978) reported that novice teachers and newly tenured teachers are often
transferred to teach subjects that lack senior teaching staff. Difficult and un-
popular academic major subjects, such as math, are the most common subjects
for low-seniority staff to receive as their teaching assignments. Green found that
the worst schools in her survey had no teachers who had been mathematics
majors teaching math to the students, whereas the best schools studied had only
teachers with college math majors teaching mathematics. The schools with the
highest math achievement scores had teachers with the greatest amount of experience in math teaching and the highest number of teachers with master's degrees
in mathematics. Thus, school systems can maximize teaching effectiveness by
providing teachers who are trained in specialized content areas, such as mathematics.

A second aspect of "opportunity to learn" that research has uncovered in
recent years is interventions that can result in higher math achievement for
bilingual students. It was reported previously that university remedial programs,
even those carried out in majority-status institutions, can be effective (Merino et
al., 1979). Green's data (1978) also indicate that minority staff members are a
big part of intervention effectiveness. Teachers' understanding of students'
cultures, languages, and values appears to influence effectiveness in conveying
academic subject matter and its relevance.

Experience with math and math teaching has been cited as important when
selecting teachers to teach mathematics. Learner experience is also an important
factor and, in fact, turns out to be the best single predictor of achievement
(Barnes & Asher, 1962). Additional math courses provide more than exposure to
advanced topics; they also provide additional experiences that consolidate and
maintain basic math skills acquired in earlier years. For a student to do well in
first-year college calculus, the evidence strongly suggests that the student should
be encouraged to take Geometry, Trigonometry, and Algebra II in high school.
An important by-product, particularly for the bilingual learner, is the additional
experience with the language of mathematics, the math symbol system, and the
mathematics-language expressions that are extended in the course of pursuing a
variety of topics in mathematics. This is the concept of "automaticity" of
functioning, which means simply that with experience and practice, one gains
facility in using information and skills.

Although it might seem desirable to extend the experiences in cross-mapping math concepts in natural language and mathematical representations to more than one language, there are those who argue that opportunities for learning are restricted for bilingual students when mathematics instruction is not in the dominant language (Ortiz-Franco, 1980). However, the data to support this assertion are limited, and the best known evidence appears to be by inference: Students perform better on achievement tests administered in their dominant language (Cossio, 1977). The question is, Do students also learn better if the instruction is in the dominant language? Studies by Begle (1975) and Denker (1977) concluded that Hispanics taught in Spanish performed as well as non-Hispanics who were taught in English. This better-than-usual result suggests that instruction in the dominant language is helpful and complements findings by Cossio (1977) and Macnamara (1966) reported in a previous section of this chapter.

A collateral issue to the language in which the student is instructed is the issue of the language in which the student's knowledge and learning are assessed. This problem crops up in many guises. For example, lower English proficiency skills may be associated with lower reading comprehension for bilingual Hispanics, causing lower test performance because of fewer problems being solved (Mestre, 1986). When the language of performance is not the dominant language, errors increase. Explanations for these errors have ranged from conceptual (vs. mathematical) by Morales et al. (1985) to speed (Mestre, 1986) and language representation (Macnamara, 1966). The point is, performance differences are more likely to appear when either assessment or teaching occurs in the nondominant language for the bilingual learner.

Finally, researchers have begun proposing models for increasing bilingual students' achievement based on the concept of "opportunities for learning." The following two examples illustrate that quite different approaches are offered to achieve the same ends when the problem has so many facets. Lopez (1979) offered a model based on an improved delivery system. He proposed that successes with bilingual students can be increased by (a) providing more and better inservice education for classroom teachers; (b) developing programs to increase parental involvement; and (c) working toward improving students' self-concepts as learners. Gallegos (1979), by contrast with Lopez, offered a model that was more directly content- and curriculum-tied. The guidelines of the Gallegos model proposed (a) developing basic vocabulary skills in English and non-English for each topic of study; (b) expanding curricula to include more theory and explanation, such as underlying mathematical rationales and mathematical theory; (c) expanding curricular topics to include less familiar content, such as geometry and probability; and (d) providing both appropriate and concrete learning materials that are designed specifically for bilingual students.

Both models are examples of attempts to improve the students' opportunities to learn. Once again, however, we have to conclude this section by stating that there is paucity of evidence for comparing the relative effects of various models or of the relative importance of any particular factor or set of factors that enhance educational opportunities for minority language students.

What is Known About the Effects of Parental Encouragement and Assistance?

When we discussed the types and extent of encouragement students receive from teachers, it was reported that majority teachers think they are protecting minority students from failure experiences by setting lower level goals than they set for majority students. We also reported that majority teachers tend to discourage minority students from pursuing the difficult and advanced mathematics courses. Minority teachers, by contrast, demonstrate another type of behavior to protect students from failure experiences: Minority teachers are more encouraging of students to pursue advanced educations, but the encouragement is not in the direction of science or mathematics careers. In a sense, then, the minority student experiences a consistent pattern of direction away from math by the teaching community. Parents add to this consistency. The Ford report (1980) indicated that parental encouragement patterns are similar to those of teachers, and for minority students this means that they are directed away from careers that are dependent upon skills of math and science.

Parental attitudes toward mathematics and mathematicians also influence achievement indirectly. Green (1978) and MacCorquodale (this volume), as reported previously, found that native Americans and Hispanic women were not encouraged in math because the discipline is associated with character qualities that are undesirable to the cultures.

Thus, whatever else the differences and inconsistencies appear to be between homes and schools for language-minority students, one message is clear and consistent: Mathematics education and mathematics careers aren't for you.

Is Mathematics Mastery Adequately Assessed in the Bilingual Learner?

Measurement issues are particularly important to the present chapter and the controversy surrounding bilingual or minority assessment, in recent years, has been substantial. The measurement issues range from fairness in testing to validity—essentially the whole assessment gamut. The National Council of Teachers of Mathematics (NCTM) pointed to four major concerns with respect to bilingual student assessment (Ortiz-Franco, 1980):

1. Tests in English, standardized or not, do not measure accurately the knowledge and academic achievement of bilingual students. Adequate testing instruments need to be developed.

2. Local school districts ought to move away from reliance upon national norms and develop local norms that are more realistic and relevant indicators of mathematics achievement in their specific school populations.

3. It is frequently the case that bilingual and minority students get low scores on tests due to unfamiliarity with necessary test-marking strategies. These students should be taught test-taking techniques so that students' actual knowledge can be assessed more accurately.

4. Item analysis, test reliability, and validity ought to be integral parts of the testing process when bilingual students are involved.

The concerns raised by NCTM, quite likely, are not unique to math mastery assessment for language minority students. These concerns are larger than the bilingual issues, but they become particularly important when assessment is in a nonnative language and for a culturally different group. Issues of coaching, test validity, adequacy of the instruments, and the appropriateness of indicators of achievement all need to be addressed for mathematics education, in general.

Ginsburg has addressed the testing issue directly by offering suggestions for how to go about revising our mastery assessment procedures, rather than leaving the suggestions at the "what needs to be done" level (1981, 1982). He suggested, at least at the lower grades in school, that interview procedures are more sensitive and appropriate methods of assessment than traditional standardized instruments. The trick is to get the teacher skilled in the procedures so that the interview becomes both diagnostic and prescriptive. The relative efficacy of these procedures vis-á-vis traditional methods, however, is still to be determined.

Comment

The problem of clarifying the meanings and the indicators of independent variables is complex. Although it is being increasingly recognized that both sociocultural and school program variables must be investigated as influences on student achievement, it is also increasingly apparent that these intervening variables offer a range of diversity that makes it difficult to equate their importance across various ethnic groups (Simmons, 1985). It is clear that these moderator variables need as much attention as the dependent variable itself. Although we seem to know which are the important influences, we do not have much information about their combined effects upon math achievement among bilingual learners.

What Happens with Bilingual Mathematics Achievement in the
Course of Development?

The various models of academic achievement among minority students (bilinguals constituting a special subgroup within the models) can all be criticized as static models. How do the various models change if they are envisaged as moving through time? Do the moderator variables, for example, have different effects depending upon the point in development of the impact? Time of impact, in this sense, includes, of course, age of the child when he or she is learning. Equally important, knowledge builds upon knowledge and skills have other skills as their foundations. As such, the developmental concerns of this analysis included (a) Does the math deficit of minority bilinguals become a cumulative deficit in bilingual secondary school learners? (b) How does language, as an associated variable of math achievement, affect abstract math learning among older students as compared with its impact on more rudimentary and concrete math skills during learning by younger students? (c) Can we regard the math/language problem as really being a different educational problem for elementary learners than for secondary students, since both math and language are more dependent upon abstract concepts and more abstract symbol systems for advanced learners?

Developmental analysis points out problems associated with long histories of low achievement, as well as the possibly changing dependencies of cognitive skills upon related and/or supportive skills, and raises the possibility that the questions themselves may change over time.

Research on aspects of these problems has brought us to a point where, fortunately, there are reasonable data from which to draw some inferences. The disappointing aspect of these studies is that, although we may make developmental inferences and draw developmental conclusions from the cross-sectional or comparative studies, the research itself has not been directly developmental or conducted within a developmental framework. What we have at our disposal are studies that demonstrate the so-called Age Differentiation Hypothesis (older children can do more things than younger children) and studies that demonstrate developmental changes.

In general, we know that math ability increases with age in normal learners during the school years. However, levels of achievement become more diverse (Jarvis, 1964). The range in mathematical ability of sixth graders may be as much as 7 years. Despite the hypothesis of an invariant sequence of developmental stages, Piaget's work acknowledged that there can be considerable variation in age of attainment of any stage. Age of attainment is sensitive to factors as pervasive as whether one is an urban or a rural dweller, irrespective of country of origin, as demonstrated by comparative studies in the United States, Iran, Australia, and West Africa (Dasen, 1972; Mohensi, 1972; Piaget & Inhelder, 1969).

Aiken (1971a) concluded that the factor structure of math abilities is more complex in high school and college students than for students in the lower grades. Thus, researchers conclude that the factor structure of math ability becomes more differentiated with age (Dye & Very, 1968; Very, 1967). Although males tend to show more sharply differentiated factors than females, the reasons are attributed to greater exposure of males to math experiences. For the purposes of this discussion, we have to consider the impact of educational opportunities and increased experiences as potential contributors to greater mathematical differentiation that occurs in the course of schooling. Although the hypothesis is termed *Age Differentiation Hypothesis,* it is just as reasonable to consider this as a broader *Developmental Differentiation Hypothesis* because research seems to indicate that the learning opportunities change during the schooling process for males and females, as well as for majority and minority students.

What developmental changes have been documented for math learning in language minority students? Begle (1975) looked at mathematics achievement in Mexican-American students. The study included both English/Spanish bilinguals and Spanish monolinguals at first, third, and ninth grades, all of whom were low SES. Bilingual program results indicated that first graders showed improved learning on basic concepts and in computational skills, whereas reasoning skills did not get any better. Third graders, by contrast, showed improved reasoning skills but no further gains in computational skills. No significant differences in mathematics achievement were noted for ninth graders after participating in a bilingual program. The suggestions of these results are that (a) the information to be learned changes as the grade level goes higher; (b) the learner's information processing strategies change with development; (c) symbolic functioning increases with development and there is less reliance on concrete features for thinking; and (d) bilingual programs have differential impact on various math skills and these impacts vary with the background skills the learner brings to the learning situation: his or her developmental status.

Thus, we know more about the impact of bilingualism upon math achievement than mere statistics that report age differences among 9-, 13-, and 17-year-olds (e.g., NAEP, 1979). Despite the limitations of research synthesis and cross-sectional studies, it can be concluded that Hispanics learn computational skills and basic concepts in first grade and reasoning skills in third grade best if taught in programs that emphasize both languages. Comparable studies for other language minority students in the United States remain to be reported in the educational research literature.

Are there any Data on Instructional Programs or Interventions with Bilingual Mathematics Learning?

Begle's literature review (1975) indicated that both computational and reasoning skills can be enhanced for Hispanic learners through bilingual instruction, although the benefits were different for the various age groups. Denker (1977) also found that Hispanics who were taught math in Spanish performed as well as non-Hispanics who were taught math in English. We can conclude, then, that there is some evidence to suggest the value of intervention programs for students who are ''at-risk'' educationally. What is not specified are the details that might account for program effectiveness and the relative merits of various intervention strategies.

Educational interventions can never be divorced from the deliverers of the programs and the resources available to the programs. Hence, the value of instructional programs is intertwined in a network of variables including teacher competencies, educational opportunities, and instructional resources.

The concerns enumerated by the National Council of Teachers of Mathematics (NCTM) underscore the point that there is more to the issue than providing math instruction in the dominant language. The concerns fall into two general classes: those dealing with curricula and those dealing with teaching (Ortiz-Franco, 1980).

I. Curricular Concerns.
  1. There are no complete, coherent, and comprehensive bilingual elementary school mathematics programs available;
  2. Many topics in mathematics are independent of language and culture so long as the fundamental preskills are adequately strengthened;
  3. The development of quality bilingual mathematics materials emphasizing learning by seeing, doing, and participating needs attention;
  4. Studies of how much mathematics knowledge and skill immigrant students have are lacking. This lack, in turn, makes decisions concerning the placement of these students in the existing local program difficult;
  5. Studies comparing the mathematics curriculum prevalent in the United States and in the country of origin of immigrant students should be undertaken.
II. Teaching and Teacher Education Concerns.
  1. When teaching mathematics to bilingual students in a bilingual atmosphere, teachers should have (a) an adequate command of the technical math vocabulary in the language involved; (b) a sound preparation in the mathematical concepts of the elementary school curriculum; (c) techniques for minimizing the effects of the language and cultural variables with the purpose of maximizing mathematical thinking and mathematical processes;
  2. More emphasis on mathematics is needed in the preservice training programs for prospective elementary school teachers in order to increase their effectiveness in teaching mathematics;
  3. Preservice mathematics education programs for teachers should (a) encourage prospective teachers to utilize homemade teaching materials; (b) facilitate first-hand experience in problem solving and applications of mathematical concepts in a variety of different situations; and (c) enable prospective teachers to develop math labs and math interest centers.
  4. Courses integrating mathematics content, information about the learning characteristics of children, and mathematics teaching methods are needed.

5. Bilingual mathematics specialists should be used at the K–6 levels to impart mathematics instruction in the vernacular language of the students.

The guidelines offered by Lopez (1979) and Gallegos (1979), all mentioned previously, have not been tested either experimentally or empirically. Although many curricular and teacher training recommendations have been and continue to be made, evidence for any of these proposals is scanty.

How General or Specific is the Mathematics Learning Problem
Among Bilinguals of Varying Ethnic Backgrounds?

The statistics that have been cited for the gap between minority and majority students in math achievement have been for a fairly restricted set of the language minority groups in the United States. Although Hispanics make up the largest population of language minority people, and will be the largest minority group in the United States by the year 2000 A.D., we try to provide effective education for all American children; hence, we must ask whether or not other non-English-dominant students in American schools do or will have the same achievement deficits. We infer that they might, based on two sources of evidence. First, the dependency of math concept learning upon language proficiency leads us to conjecture that the problem is general to what happens when one is learning in one's nondominant language. Second, low math achievement in minority language groups other than Hispanics is evidence of the widespread nature of the problem. Native Americans, for example, have not achieved at levels comparable to majority students (Green, 1978; Leap, this volume).

What comes to mind in discussing these statistics, however, is the fact that Asian-American students in the past have generally demonstrated high math achievement, certainly higher than any other minority group. Does this mean, first, that Asian-Americans do not experience any difficulties coping with mathematics; or second, that the problem is specific to certain language communities? Some have reported that Asian-Americans have experienced the reverse of other language minorities: Asians are denied many educational opportunities because they are encouraged only in areas that require mathematical performance (Chan & So, 1982; Chun, 1980; Tsang, 1984).

Only one comparative study can offer any insight into the question of the breadth of between and within ethno-linguistic group aspects of the problem. De Avila & Duncan (1980) looked at school achievement in nine ethno-linguistic groups of subjects:

1–2 Urban and Rural Mexican-Americans
  3 Puerto Rican
  4 Cuban American

5 Chinese
6 French Cajun
7 Native Americans
8 Anglo
9 Mexican - Mexico City

DeAvila and Duncan's analyses looked at the interactions of ethnic group membership, sex, and grade level, using language proficiency as the dependent variable. Although 10% of the variance was attributed to gender differences, ethno-linguistic group membership accounted for 39% of the total variance. As would be expected, Anglo children were the most proficient in English, followed by Cuban and French background children. The conclusion was that language proficiency varies greatly across various ethno-linguistic groups. We might conclude that all groups do not face the same problems and even if they do, the extent of ethnic diversity indicates that the same remedies may not be appropriate for all groups.

We are confronted with another set of statistics that have nothing to do directly with math achievement or language proficiency. The influx of immigrants in the past few years has meant that the education population is changing. Previous findings indicating high scholastic achievement among Asian-American students may no longer hold, now that the number include greater within-Asian group diversity. Other immigrant groups have increased, as well, and we do not know what to expect with regard to their school success. Thus, where we previously based our generalizations on between-ethnic-group differences, such as Asian and Hispanic comparisons, the new statistics may indicate greater within-group than between-group differences, such as difference between Chinese and Vietnamese or among Cubans, Chicanos, and Puerto Ricans.

Part of the problem in predicting academic achievement for new immigrants involves a problem with which we have not grappled successfully for our own culture: social class. Apart from ethnic diversity, we are now receiving waves of immigrants who are quite different in social class origin from their predecessors, and as a result we have poor indicators to use in predicting their success in school. Whereas Vietnamese immigrants of the 1960s were entrepreneurial class or spouses of Americans, later immigrants were refugee boat-people from the war-torn provinces. Cuban immigrants of 20 years ago were also business people, by and large; more recent immigrants have been very poor and poorly educated.

Puerto Rican students provide a contrast to all other immigrant groups because of the unique territorial status of the island. If one does not see oneself as an immigrant, or if one anticipates returning to Puerto Rico, is there likely to be strong motivation to merge into the educational mainstream or to learn English (Ortiz, 1983)? Vietnamese and Cubans know with greater certainty where they will call home for the next few years.

CONCLUDING REMARKS

New questions must be formulated in order to understand the language-academic-achievement relationships. Where we previously asked about the language-achievement relationship directly for a specific ethnic group, we now have to ask about generational differences and about differences in the reasons for, and time of, immigration. Thus, ethnic group membership questions will now be seen in interaction with generational and/or period of immigration. These, in turn, interact with the variable that has always been the social scientist's central concept: social class.

The model that will emerge in our new conceptualization of the language-achievement relationships will contain the same entries as outlined previously, but the meanings of those entries seem to be changing: the intervening variables are becoming increasingly differentiated as the groups whom the data represent become more varied. Any graphic representation of the model should also have some horizontal arrows to depict change within the model itself as a function of time. Psychosocial, Sociocultural, Pedagogical, Cognitive, and Linguistic factors remain relevant, but time has changed the meaning of the relationships that were reported in the small research studies of several decades ago.

The research questions, the social class variables, the learning patterns—in fact, the basic assumptions about language and learning—seem to be changing as the educational population becomes more ethnically diverse. It is time to reexamine previously held assumptions and then ask what else we need to know in order to meet the challenges of educational quality and educational equity.

REFERENCES

Aiken, L. (1971a). Verbal factors and mathematics learning: A review of research. *Journal for Research in Mathematics Education, 2,* 304–313.

Aiken, L. (1971b). Intellective variables and mathematics achievement: Directions for research. *Journal of School Psychology, 9,* 201–212.

Barnes, W., & Asher, J. (1962). Predicting students' success in first year algebra. *Mathematics Teacher, 55,* 651–654.

Begle, E. G. (1975, March). *Test factors, instructional programs, and socio-cultural economic factors related to mathematics achievement of Chicano students: A review of the literature.* Palo Alto, CA: Stanford Mathematics Education Group, Stanford University.

Bruner, J. S. (1975). Poverty and childhood. *Oxford Review of Education, 1,* 31–50.

Bruner, J. S., & Kenney, H. J. (1965). Representation and mathematics learning. Mathematical learning. *Monographs of the Society for Research in Child Development, 30*(1, Serial No. 99).

Call, R., & Wiggen, N. (1966). Reading and mathematics. *Mathematics Teacher, 59,* 149–57.

Chan, K., & So, A. Y. (1982). *The impact of language of instruction on the educational achievement of Hispanic students* (Technical Note #8). Los Alamitos, CA: National Center for Bilingual Research.

Chipman, S. F., & Thomas, V. G. (in press). The participation of women and minorities in mathematical, scientific, and technical fields. *Review of Research in Education, 14.*

43

Chipman, S. F., Brush, L., & Wilson, D. (1985). *Women in mathematics: Balancing the equation.* Hillsdale, NJ: Lawrence Erlbaum Associates.

Chun, K. (1980). The myth of Asian American success and its educational ramifications. *IRCD Bulletin* XV (1 & 2). NY: Teachers College, Columbia University.

Cleland, D., & Toussaint, I. (1962). Interrelationships of reading, listening, arithmetic computation, and intelligence. *Reading Teacher, 15,* 228–231.

Clement, J., Lochhead, J., & Monk, G. (1981). Translation difficulties in learning mathematics. *American Mathematical Monthly, 4,* 286–290.

Cocking, R. R., & Potts, M. (1977). Social facilitation of language acquisition: The reversible passive construction. *Genetic Psychology Monographs, 94,* 249–340.

Cossio, M. (1977). The effects of language on mathematics placement scores in metropolitan colleges. *Dissertation Abstracts International, 38*(7), 4002A.

Cottrell, R. (1968). A study of selected language factors associated with arithmetic achievement of third grade students. *Dissertation Abstracts International, 68,* 5505.

Dasen, P. (1972). Cross-cultural Piagetian research: A summary. *Journal of Cross-Cultural Psychology, 3,* 23–40.

De Avila, E. (1980). *Relative language proficiency types: A comparison of prevalence, achievement level, and socioeconomic status.* Report submitted to the RAND Corporation.

De Avila, E., & Duncan, S. (1980). *The language minority child: A psychological, linguistic, and social analysis* (Report No. 400-65-0051). Washington, DC: National Institute of Education.

Denker, E. (1977, April). *Teaching numeric concepts to Spanish-speaking second graders: English or Spanish instruction?* Paper presented at American Educational Research Association, New York.

Duran, R. (1985). Influences of language skills on bilinguals' problem solving. In S. F. Chipman, J. Segal, & Glaser, R. (Eds.), *Thinking and learning skills* (Vol. 2, pp. 187–202). Hillsdale, NJ: Lawrence Erlbaum Associates.

Dye. N., & Very, P. (1968). Growth changes in factorial structure by age and sex. *Genetic Psychology Monographs, 78,* 55–58.

Erickson, L. (1958). Certain ability factors and their effect on arithmetic achievement. *Arithmetic Teacher, 5,* 287–293.

Feierabend, R. (1960). *Review of research on psychological problems in mathematics education.* In Research problems in mathematics education, Monograph #3. U.S. Office of Education, 3–46.

Gallegos, T. (1979). A methods course for the bilingual elementary school mathematics teacher. *Dissertation Abstracts International, 39*(11), 6603A.

Gilmary, S. (1967). Transfer effects of reading remediation to arithmetic computation when intelligence is controlled and all other school factors are eliminated. *Arithmetic Teacher, 14,* 17–20.

Ginsburg, H. (1981). Social class and racial influences on early mathematical thinking. *Monographs of the Society for Research in Child Development, 46* (6, Serial No. 193).

Ginsburg, H. (1982, June). *The clinical assessment of competence.* Address at the 12th Symposium of the Jean Piaget Society, Philadelphia, PA.

Goodrun, L. (1978). Relationship of differences between arithmetic computation and word problem solving scores and teacher ratings of oral language performance for fourth graders. *Dissertation Abstracts International, 39*(5), 2801A.

Green, R. (1978). Math avoidance: A barrier to American Indian science education and science careers. *BIA Educational Research Bulletin, 6,* 1–8.

High school and Beyond Language File Code Book. (1978). Washington, DC: National Center for Educational Statistics.

Jarvis, O. (1964). An analysis of individual differences in arithmetic. *Arithmetic Teacher, 11,* 471–473.

Kaput, J., & Clement, J. (1979). Letter to the editor. *The Journal of Children's Mathematical Behavior, 2*(2), p. 208.

Lambert, W. (1972). *Language, psychology and culture.* Stanford, CA: Stanford University Press.

Lochhead, J. (1980). Faculty interpretations of simple algebraic statements: The professor's side of the equation. *Journal of Mathematical Behavior, 3,* 29–37.

Lochhead, J., & Mestre, J. (in press). The language of algebra. *1988 Yearbook of the National Council of Teachers of Mathematics.* Rosslyn, VA: NCTM.

Lopez, A. (1979). Instructional needs in schools with significant Spanish-speaking as perceived by teachers and administrators. *Dissertation Abstracts International, 39*(9), 5311A.

Macnamara, J. (1966). *Bilingualism in primary education.* Edinburgh: Edinburgh University Press.

Madden, R. (1966). New directions in the measurement of mathematics ability. *Arithmetic Teacher, 13,* 375–379.

Merino, B. (1983). Language loss in bilingual Chicano children. *Journal of Applied Developmental Psychology, 4,* 277–294.

Merino, B., Politzer, R., & Ramierz, A. (1979). The relationship of teachers' Spanish proficiency to pupils' achievement. *Journal of the National Association for Bilingual Education, 3,* 21–37.

Mestre, J. P. (1981). Predicting academic achievement among bilingual Hispanic college technical students. *Educational and Psychological Measurement, 41,* 1255–1264.

Mestre, J. (1986). Teaching problem-solving strategies to bilingual students: What do research results tell us? *International Journal of Mathematics Education, Science & Technology, 17,* 393–401.

Mestre, J., & Lochhead, J. (1983). The variable-reversal error among five cultured groups. In J. Bergeron & N. Herscovics (Eds.), *Proceedings of the Fifth Annual Meeting of the North American Chapter of the International Group for the Psychology of Mathematics Education* (pp. 180–188). Montreal, Canada: International Group for the Psychology of Mathematics Education.

Mestre, J., & Robinson, H. (1983). Academic, socioeconomic, and motivational characteristics of Hispanic college students enrolled in technical programs. *Vocational Guidance Quarterly, 31,* 187–194.

Mohensi, N. (1972). A study of children schooled in Teheran. In J. Piaget (1973). *The child and reality.* New York: Grossman.

Moll, L. M., Estrada, E. J., Diaz, S., & Lopes, L. (1980, July). *The organization of bilingual lessons: Implications for schooling.* The quarterly newsletter of the Laboratory of Comparative Human Cognition, 2(3), 53–58.

Morales, R. V., Shute, V. J., & Pellegrino, J. W. (1985). Developmental differences in understanding and solving simple mathematics word problems. *Cognition & Instruction, 2,* 41–57.

Morning, C., Mullins, R., & Penick, B. (1980). *Factors affecting the participation and performance of minorities in mathematics.* New York: Ford Foundation.

National assessment of educational progress: Changes in mathematical achievement, 1973–78. Washington, DC: Government Printing Office, 1979.

National Assessment of Educational Progress (NAEP), Annual Report. Washington, DC: Government Printing Office (1980).

Ortiz, V. (1983). Generational status, family background, and educational attainment among Hispanic youth and non-Hispanic white youth. In M. Olivas (Ed.), *Latino college students* (pp. 29–46). New York: Teachers College Press.

Ortiz-Franco, L. (1980). *First glances at language and culture in mathematics education.* Unpublished manuscript, National Institute of Education, Washington, DC.

Piaget, J. (1973). *The child and reality.* New York: Grossman.

Piaget, J., & Inhelder, B. (1969). *The psychology of the child.* New York: Basic Books.

Pitts, R. (1952). Relationships between functional competence in mathematics and reading grade levels, mental ability, and age. *Journal of Educational Psychology, 43,* 486–92.

Potts, M., Carlson, P., Cocking, R., & Copple, C. (1979). *Structure and development in child language.* Ithaca, NY: Cornell University Press.

Rose, A., & Rose, H. (1961). Intelligence, sibling position, and sociocultural background as factors in arithmetic achievement. *Arithmetic Teacher, 8,* 50–56.

Rosnick, P. (1981). Some misconceptions concerning the concept of variable. *The Mathematics Teacher, 74,* 418–420.

Rosnick, P., & Clement, J. (1980). Learning without understanding: The effect of tutoring strategies on algebra misconceptions. *Journal of Mathematical Behavior, 3,* 3–27.

Simmons, W. (1985). Social class and ethnic differences in cognition: A cultural practice perspective. In S. F. Chipman, J. Segal, & R. Glaser, *Thinking and learning skills* (Vol. 2, pp. 519–536). Hillsdale, NJ: Lawrence Erlbaum Associates.

Spencer, P. L., & Russell, D. (1960). Reading in arithmetic. In F. E. Grossnickle (Ed.), *Instruction in arithmetic* (pp. 202–223). Twenty-Fifth Yearbook of the National Council of Teachers of Mathematics. Washington, DC: NCTM.

Tsang, S. L. (1984). The mathematics education of Asian Americans. *Journal for Research in Mathematics Education, 15,* 114–122.

Very, P. (1967). Differential factor structures in mathematics ability. *Genetic Psychology Monographs, 75,* 169–207.

Williams, F. (1970). *Language and poverty.* Chicago, IL: Markham.

Wise, L. L. (1985). Project TALENT: Mathematics course participation in the 1960s and its career consequences. In S. F. Chipman, L. Brush, & D. Wilson, *Women and mathematics: Balancing the equation.* Hillsdale, NJ: Lawrence Erlbaum Associates.

Wrigley, F. (1958). Factorial nature of ability in elementary mathematics. *British Journal of Educational Psychology, 28,* 61–78.

Zimiles, H. (1976). *Young Native Americans and their families: Educational needs, assessments, and recommendations.* New York: Bank Street College of Education.

Chapter 3

Linking Language with Mathematics Achievement:
Problems and Prospects

Geoffrey B. Saxe
Graduate School of Education
University of California, Los Angeles

INTRODUCTION

This chapter presents a conceptual analysis of studies that relate language background to mathematics achievement. My concern is first to distinguish between models that posit direct or *intrinsic* effects of language background on mathematics achievement from models that posit indirect or *extrinsic* effects and then to discuss problems and prospects for discovering intrinsic effects. I end with a discussion of *culture* as a source of extrinsic effects and examine how cultural background interweaves with children's developing mathematical cognitions in two distinct settings.

In order to distinguish between *extrinsic* and *intrinsic* effects clearly, it is useful to consider briefly some aspects of the article by Cocking and Chipman in this volume. Cocking and Chipman outline classes of extrinsic factors that relate language background to mathematics achievement. These include the following:

Entry Mastery. Children enter school with different degrees of language competence. This could influence how much children benefit from mathematics instruction.

Opportunities to Learn. Bilingual mathematics teachers, due to their short supply, may be selected for their bilingual skills, rather than their competence in teaching mathematics. Math may have low priority in programs for language minority students. Language minority students may systematically differ from nonlanguage minority students in the amount of access to parents or others who can give them tutoring and enriched educational experience.

Motivational Factors. Language minority children may be exposed to norms that systematically differ from those of nonlanguage minority children, norms that differ with respect to the value placed on mathematics education. The mathematics curricula may differ in relevance to the language minority and nonlanguage minority student.

Measurement Factors. The language minority student may be handicapped with respect to how mathematics achievement is assessed. Word problems may cause special problems for the language minority student. Tests may be biased toward sensitivity to the mainstream culture.

These factors are extrinsic in the sense that they happen to be part of the bilingual child's predicament at this historical point of our educational system and not inherently linked to growing up bilingual. In contrast to these extrinsic factors, it is also plausible that there are aspects of bilingualism that inherently influence children's mathematical cognitions. By focusing the larger part of this chapter on the intrinsic influences, I do not mean to minimize the importance of the extrinsic factors. In fact, it is the extrinsic factors that are probably the most powerful ones today mediating the relation between language background and mathematics achievement. However, if we are to work toward a psychological understanding of the influence of language background on mathematics achievement, any account must face squarely the problem of whether there are intrinsic factors, and, if there are, how to proceed in studying them.

AN ANALYSIS OF SOME INTRINSIC FACTORS

The point of departure for this discussion is two studies (Ben-Zeev, 1977; Ianco-Worrall, 1972) that offer support for the argument that there are intrinsic relations between language background and certain aspects of cognitive development. Using these studies as a base, I argue that a case can be made for the possibility of finding intrinsic relations between language background and mathematical cognition. Ben-Zeev and Ianco-Worrall each ground their research in aspects of Piaget's (1929) and Vygotsky's (1962) formulations of cognitive development: During the early phases of language acquisition, children believe that words are not arbitrary conventions to signify objects; with development they come to understand the conventional origins of words. Both Ben-Zeev and Ianco-Worrall argue that because bilingual children must sort out two separate systems of signification and syntax that are functionally equivalent, these children should develop an early understanding of the arbitrary and functional properties of language. I briefly review the studies here to give an overview of the methods and findings, and then point out some implications of these investigations for our understanding of developmental processes in children's formation of certain kinds of mathematical concepts.

Review of the Two Studies

The Ianco-Worrall Study

The setting of Ianco-Worrall's (1972) study was in Pretoria, South Africa. The sample consisted of 30 Afrikaans-English bilinguals at each of two age levels, 4- to 6-year-olds and 7- to 9-year-olds and monolingual controls.

Ianco-Worrall used two measures as dependent variables to assess children's awareness of the relation between a word and its referent. The first was used to determine whether bilinguals were more attuned to the semantic as opposed to the phonetic similarities of words at a younger age than monolingual children (Semantic and Phonetic Preference Test). For example, the child was told: "I have three words: cap, can, and hat. Which is more like cap, can or hat?" The second measure was an adaptation of Vygotsky's interview technique to assess children's understanding of the arbitrary relation between a word and its referent. The measure included assessments of children's understanding that the names of objects are interchangeable both in and out of play contexts.[1]

Ianco-Worrall found partial support for her thesis concerning the facilitative effects of bilingualism. In the younger (but not the older) age groups, bilinguals consistently chose the word that was semantically rather than phonetically similar. On the Vygotskiian procedures, bilinguals more frequently judged that objects could be interchanged in the "out of play" (but not the "in play") measures.

The Ben-Zeev Study

Ben-Zeev also found support for the thesis that bilingualism interacts with some conceptual aspects of language development. The setting of Ben-Zeev's (1977) study was in the United States (Chicago and Brooklyn) and Israel (Jerusalem and Tel Aviv). The sample consisted of ninety-five 5- and 8-year-olds divided into four groups: Hebrew-English bilinguals (United States), Hebrew-English bilinguals (Israel), English monolinguals (United States), Hebrew monolinguals (Israel). Ben-Zeev also included assessments for children's understanding of the arbitrary properties of language. For instance, in her Symbol Substitution Test, the child was required to substitute one meaningful word for another, usually within the context of a sentence. The earlier items in the battery required the substitution of one noun for another noun. For example, the interviewer would say, "You know that in English this is named airplane (experimenter shows toy airplane). In this game its name is turtle . . . . Can the turtle fly?

---

[1]Out of play context: Children were asked whether given names could be interchanged. For instance, the child was asked "Suppose you were making up names for things, could you then call a cow 'dog' and a dog 'cow'?" In play context: The child was told something like the following. "Let us play a game. Let us call a dog 'cow.'" This was then followed by two questions: "Does this 'cow' have horns?" and "Does this 'cow' give milk?" This was repeated using two other examples.

(correct answer: yes) How does the turtle fly? (correct answer: With its wings)." Later items, however, required the substitution word to violate obligatory selection rules. For example, the interviewer would say, "For this game the way we say I is to say macaroni. So how do we say 'I am warm'?" (Correct answer: Macaroni am warm; incorrect answer: Macaroni is warm.) Finally, one item required the substitution of a major part of speech for a minor one: the word *clean* was substituted for the word *into,* as in "The doll is going to clean the house." Ben-Zeev (1977) points out that, unlike the measure used in the Ianco-Worrall study,

> Success on the later part of the test implies something more than simply differentiating a word from its referent. . . . For these items it is necessary to ignore not only the semantic meaning of individual words but also the selectional rules which govern the usual relationships between classes of words within the sentence . . . treating the sentence as an arbitrary abstract code. Having experienced more than one language code system, the bilingual should be freer to abandon the rules of a particular language system for a different set of rules when this is necessary. (p. 1012)

Ben-Zeev found, as predicted, that bilinguals performed significantly better on the symbol substitution measures than monolinguals, despite the finding that the monolonguals achieved higher vocabulary scores on the WISC-R subtest, a widely used psychometric measure of intelligence.

Some Notes on Methods

The two studies reviewed share a number of virtues. Each study is embedded in a normative conceptualization of a cognitive developmental process (children's developing understanding of the arbitrary properties of language), and each author offers an articulated formulation about how and why bilingualism should interact with this normative process. Despite the merits of the Ianco-Worrall and Ben-Zeev studies, the studies nonetheless show some general methodological problems that are endemic to any study on intrinsic effects of language background on cognitive development. It is useful to review some of these problems, since they bear on research linking language background with mathematics achievement.

The ex post Facto Design.   A problem endemic to any study on the effects of bilingualism on cognitive development is that the assignment of the subjects to experimental and control groups is not random. Rather, there are a host of factors that are systematically related to being "bilingual" as opposed to "monolingual." As a consequence, if significant differences are obtained between bilingual and monolingual samples, one can never know if one has found a factor that is *intrinsic* to the process of growing up bilingual (as may be the want of the researcher) or *extrinsic* to the process (an uncontrolled factor systematically associated with bilingualism).

There are two typical research strategies used to address this problem of controlling for the systematic influence of extraneous variables. Both require estimates of the plausible extraneous variables. Then, the researcher either matches the bilinguals and monolinguals on the measured dimensions or statistically controls for these dimensions in the analysis of subjects' performances. Both of the studies reported above used one or the other of these techniques. Ianco-Worrall matched children on several dimensions (e.g., intelligence) whereas Ben-Zeev controlled statistically for several possible extrinsic factors. These solutions to the problem of nonrandom assignment of subjects to population groups are not completely satisfactory. A researcher may be unaware of the parameters associated with the bilingual experience that could conceivably influence performance on the measures of cognition, and consequently he or she may have simply not included an estimate of the factor that may actually produce observed differences in cognitive performances. The fact that Ianco-Worrall and Ben-Zeev controlled for overlapping (but not identical) factors points to the risk in this kind of design.

Controlling for Intelligence.   In studies on intrinsic effects, it is often expected that the bilingual population will outperform monolinguals on a dimension that is closely linked to (if not a defining characteristic of) intelligence (e.g., understanding the arbitrary properties of names). If the variation is reduced on the dependent measure (e.g., estimates of cognitive development) by controlling for IQ, then the possibility of finding a difference between population groups, if one actually exists, is considerably reduced (since one is equating subjects on a factor that may be a defining characteristic of the dependent measure). For instance, on the "in play" measure of the Ianco-Worrall study, no differences were found between the bilingual and monolingual groups. This may have been because there were no differences. Alternatively, there may have been differences, but, since the variability on this measure may have been reduced by matching monolingual and bilingual children on an aspect of intelligence also affected by growing up bilingual, systematic differences between monolingual and bilingual children's understandings may have been obscured. The problem has no clear solution since the alternative is no more attractive. If intelligence is left free to vary across groups, it is always possible that it is systematically related to population group for reasons other than an intrinsic relation between population group and language background. A better alternative to controlling for IQ would be one dictated by a comprehensive conceptual model. In this case, theory ought to specify how a particular operational definition of intelligence ought to be related to performance on some dependent measures assessing the influence of language background. If this condition were met, we would have a less formidable task of making sense out of relations between language background, measures of intelligence, and performance on our dependent measures of cognition. I offer one approximation to this approach later in the studies linking language background to mathematical cognition.

Defining Language Background. A third problem is the criteria researchers use to define *bilingualism* and *monolingualism*. Ways in which these terms have been operationalized in the past include reference to teacher reports, vocabulary tests in the two targeted languages, tests of translation from one language to another, and standardized language assessment tests (e.g., the Language Assessment Battery). Moreover, the conceptual definitions for monolingualism and bilingualism also differ considerably. These include dimensionalizations such as restricted/elaborated codes, comprehension skills, and production skills. If the generalizations are to be made from studies linking language background, some general specification about how language background is defined is needed. (See article by De Avila in this volume for a further discussion of this topic.)

EXTENDING RESEARCH ON INTRINSIC EFFECTS
TO MATHEMATICS ACHIEVEMENT

In the first part of this section, I argue that mathematics displays properties similar to language on precisely those dimensions that Ianco-Worrall and Ben-Zeev found a positive influence for bilingualism. In the second part of this section, I review some recent studies of my own that offer a method of studying intrinsic relations between language background and mathematics achievement.

Parallels Between Language and Numeration Systems

The arbitrary properties of language that Ianco-Worrall and Ben-Zeev point to are not unique to language but are defining characteristics of any notational system. Similar to language, numeration systems consist of two types of arbitrary features. First, any numeration system consists of arbitrary markers or tags (numerals) that are used to refer to particular conceptual objects. For instance, the particular list of markers "1,2,3,4, . . .n" refers to summations of either ordinal or cardinal correspondence relations, or *numbers*. There is nothing inherent in these markers that render them *numerals*. Similar to word-referent relations, the link between a numeral and a number is arbitrary and a matter of cultural convention. Second, the base-structure rules that define the organization of a numeration system and that are a basis for arithmetical algorithms have conventional properties. For instance, though we use a base-10 numeration system (as does most of the world), there is nothing intrinsic to number that requires a base-10 system. In fact, different cultural groups use systems of different base structures, and, in our own culture, we employ numeration systems that differ from base-10. Consider the binary system used in digital computers (0 = 0, 1 = 1, 10 = 2, 11 = 3, etc.), the calendar, the quantification of eggs, and so forth.

Given the findings that bilingual children have "meta-linguistic awareness" of word-referent relations and syntactic conventions (Ianco-Worrall and Ben-Zeez) as well as the parallels I have pointed to between language and numeration, there is good reason to suspect that there may be an intrinsic relation between the process of acquiring two languages and the child's understanding of certain fundamental mathematical concepts. In the section to follow, I address the problem of investigating this relation with respect to children's understanding of number symbol-referent relations entailed in the use of number words and base structure conventions.

Some Recent Studies and Observations on Intrinsic Influences of Language Background on Mathematical Cognition

The foregoing discussion points to the need for two kinds of investigations in the study of intrinsic effects of bilingualism on mathematical achievement. The first is to identify developmental processes in mathematical cognition that may interact with growing up bilingual. It would only be once such a normative developmental analysis was completed that it would be possible to begin the second type of investigation: an analysis of the way growing up bilingual interacts with the normative developmental process. The descriptions of the observations and studies below are offered as a line of research in progress that is addressed to both of these issues.

Developmental Differences in Children's Understanding of Conventional Properties of Counting Words

The research literature on developmental changes in children's understanding of the arbitrary properties of number words is somewhat of a puzzle. Based on the literature on nominal realism (Vygotsky, Piaget), it is certainly plausible that young children treat number words with realistic properties. However, Gelman and Gallistel (1978) have reported some counterevidence to this assumption. They compared the way children assign number words to objects and the way children assign words to objects. Gelman and Gallistel found that once children assign number words to objects during a count, they generally do not have conceptual difficulty reassigning these same words to the objects in a different order. Moreover, Gelman and Gallistel also report that some children do not use the conventional list in counting. Rather, they use stably ordered but nonconventional lists of number words (e.g., 1, 2, 4, 9). On the basis of these data, then, it might be argued that ostensibly children treat number words with arbitrary properties: Children not only assign numerals flexibly to objects, but they also use stable lists other than our conventional one. However, whether this "knowledge-in-action" actually reflects an understanding of conventional properties of numeration is not altogether clear.

A Normative Study on Children's Understanding of Arbitrary Properties of Counting Words. In order to discover whether there are, in fact, conceptual changes in children's understanding of arbitrary properties of number words, I conducted a set of studies (Saxe, Sadeghpour, & Sicilian, 1986). In the first study, the focus was on the normative question: Is there a transformation in development in children's understanding of number words? To address this question, a task paradigm was devised that required children to make explicit their conceptions about the arbitrary properties of number words. Children between 5 and 11 years of age were required to evaluate the relative adequacy of the counting of two finger puppets; one puppet violated a conventional feature of counting by using the alphabet instead of standard number words to count sets of objects, whereas the other puppet violated a logical feature by establishing many-to-one instead of one-to-one correspondence between number words and objects. The child was presented with the Bert and Ernie finger puppets and told, "This is really Ernie-1 and this is really Bert-A (the interviewer turned the puppets around to show the child their labels, and then continued the story). Ernie-1 comes from down the street where they use numbers to count, but Bert-A comes from another country far away where they use letters to count." The interviewer questioned the child in order to determine if the child understood the introductory story. If the child did not understand the introduction, it was repeated and the interviewer checked the child's comprehension again. The interviewer then told the child the following: "One day Ernie-1 and Bert-A were walking along. They saw a store and went in." The interviewer presented a picture of three apples, and asked the child to identify them. Ernie-1 then says, "I wonder how many there are? I'll count. There are 1, 2, 3 (counts with his nose). There are 4." Bert-A says, "No, that's not right, that's not right at all. There are A, B, C (counts with his nose). There are C." The interviewer then asked the child, "What do you think? Did both Bert-A and Ernie-1 count right, or did one count right and the other count wrong, or did they both count wrong?" The order in which these alternatives were asked were randomized across subjects on each of the conditions. After the child responded the interviewer asked, "How do you know?" and additional probe questions. In another task condition, the interviewer told the child that Ernie-1 and Bert-A walked further on in the store. The interviewer presented a picture of three oranges, and asked the child to identify them. Ernie-1 then says, "I wonder how many there are? I'm going to count them. There are 1, 2. There are 2." Bert-A says, "No, that's not right, that's not right at all. There are A, B, C. There are C." The interviewer then asked, "What do you think? Did both Bert-A and Ernie-1 count right, or did one count right and the other count wrong or did they both count wrong?'' After the child responded, the interviewer asked "How do you know?'' In the numeral omission condition, the interviewer told the child, "Ernie-1 and Bert-A walked further on in the store.'' The interviewer presented the picture of the three pears and asked the child to identify them. Ernie-1 then says, "I wonder how many there are? I'm going to count them. There are 1, 2 (skipping the center pear). There are 2.'' Bert-A says, "No, that's not right at all. There are A, B, C. There are C.'' The interviewer then asked, "What do you think? Did both Bert-A and Ernie-1 count right, or did one count right and the other wrong or did they both count wrong?'' After the child responded, the interviewer asked, "How do you know?''

The findings demonstrated a developmental shift in children's understanding of the arbitrary property of conventions used in counting. Younger children tended not to understand the arbitrary nature of conventional symbols in that they judged the letter-counter as incorrect. In contrast, older children tend to demonstrate an understanding of the arbitrariness of conventional symbols by judging the letter counter as correct. Thus, although young children may demonstrate some kind of precursory behaviors that prefigure a later explicit awareness of conventional properties of numeration, such precursory behaviors do not imply an explicit understanding of arbitrary properties of numeration.

The counting task described above required children to know that, due to the arbitrary property of number terms, a string that normally serves a noncounting function (the alphabet) can be employed to serve a counting function. Given the Gelman and Gallistel findings, we reasoned that we might be able to demonstrate an earlier awareness of the arbitrary properties of numerical conventions if we modified our tasks. If the English letters that Bert-A counted with were substituted for a set of unfamiliar terms whose function is unknown to the children, we reasoned that children with a nascent understanding of the arbitrary property of number symbols should demonstrate a more sophisticated performance in this "unfamiliar'' condition, and this is in fact what we found. When Bert used Chinese numerals to count, monolingual American first graders demonstrated better performance than when Bert used letters to count.

An Interface Between Language Background and Children's Developing Understanding of the Arbitrary Properties of Number Words.   The next concern was to determine whether bilingual children tend to differentiate the arbitrary character of number symbols prior to their monolingual peers, as Ianco-Worrall and Ben-Zeev found with respect to language. In order to evaluate this hypothesis, we interviewed two groups of first-grade children from middle-class communities. The monolinguals were English speakers, and the bilinguals were French/English speakers from the New York City area. On the basis of teacher interview and school records, we determined that all "bilingual'' children were fluent in a language other than English (generally French). Proficiency in English was assessed by teacher interview and a test of language comprehension that was built into our assessment instrument. We regarded these language assessment procedures as appropriate since they assured some degree of fluency in both languages. In order to control for possible extrinsic factors that might lead to differential performance of our groups, our first concern was to match monolingual and bilingual children on a dimension of mathematical cognition that was fundamental to mathematical understanding, but a domain that did not require an understanding of mathematical symbolization (our dependent variable). Piaget's number conservation task was selected. It was in this way that we avoided the problems inherent in controlling for intelligence with estimates of IQ. We then assessed children's performance on the alphabet version of the Bert and Ernie tasks. As predicted, the bilinguals more frequently demonstrated an understanding of the conventional properties of number symbols than their monolingual peers.

The studies I conducted that showed the facilitative effects of bilingualism and of unfamiliar symbol sets provide empirical support for the same thesis: Underlying the child's understanding of the arbitrary property of numerical conventions is the child's cognitive dissociation of the figurative characteristics of numerical conventions from the notational function that they serve. In one part of the research described, we manipulated a task variable that reduced the child's need to disassociate the symbols for number from the notational function that they serve in everyday life (spelling) and found this manipulation facilitated children's performance. In another part of the research, we manipulated a subject characteristic, our bilingual-monolingual comparison. We found that this manipulation accomplished the same end as the task manipulation: It facilitated children's dissociation of the figurative characteristics of numerical conventions from the logical aspects of the counting process.

Developmental Differences in Children's Understanding of Base Structure Conventions

Despite the wealth of research on numerical cognition, very little research has been directed toward understanding developmental changes in children's understanding of numeration concepts. The focus of this phase of the research was on one aspect of children's developing knowledge about numeration—the strategies children use to compare a representation of one base system with a representation of another base system. The same research approach was deployed in the present studies as in the previous ones reviewed. The first concern was to understand normative developmental processes in children's cognitions about base structure principles. The second concern was to understand how growing up bilingual may interact with this developmental process.

A Normative Study on Children's Understanding of Base-Structure Conventions.   Learning to compare number representations from two different number systems is potentially confusing for children. For instance, in a base-2 system, the terms "100" and "10" differ from one another by a value of 2, whereas in a base-10 system, these terms differ from one another by a value of 90. Thus, in order to make a comparison across systems, the child must ignore the figurative properties of the symbols and translate a representation of one system into its value in the other. Such a comparison requires the tacit awareness that number representations are generated by rule systems with arbitrary properties.

In order to study developmental changes in children's comparisons of values across base systems, a task paradigm was created that was designed to sustain children's interest and not rely directly on specific knowledge about the Western numeration system. Children were presented with a puppet show in which each of two puppets used different monetary systems. The material used to represent currency was identical for each puppet, although the relative values of this material varied across different systems. For instance, in the first puppet's system, one big disk might be worth three little disks, whereas in the second puppet's system, one little disk might be worth three big disks. In the context of the puppet show, children were taught the different rule systems, and visual supports were available to remind them of the rule systems throughout the duration of the show. A comprehension check was also administered to ensure that all children knew the rule systems prior to the data collection procedure. Once the comprehension check was completed, one puppet brought money to the other puppet's country (represented on a large board) and asked for change. The other puppet gave some change, and the two puppets were made to disagree about whether the amount given was the correct amount of change. The child was then asked to intervene in the argument and to mediate the dispute by explaining whether the amount was correct and why it was or was not correct. The study made use of one task in which children were presented with currency with which they had everyday experience (pennies and nickels) and another in which they had no experience (plastic chips of two different sizes). The behavior of interest was whether and in what way children made reference to the rule systems in determining the correct exchange.

In the study, children were administered four versions of the disk and coin tasks. In two versions, the system in which the exchange was conducted was consistent with either current cultural practice (5 cents = 1 nickel) or with a plausible intuitive expectation (3 little chips = 1 big chip), and in the other two versions, the exchange was conducted in a system that was not consistent with our own cultural practice (3 cents = 1 nickel) or an intuitive expectation (3 big disks = 1 little disk). For each of these situations, on one occasion a puppet offered the correct exchange under the appropriate convention, and on the other occasion a puppet offered an incorrect exchange under the appropriate convention.

Children's responses to each of the four conditions were coded as one of three levels. Children were categorized at level 1 who made no reference to the rule systems in justifying their judgments about the accuracy or inaccuracy of the exchange. These children typically argued that the criterion for accuracy should be a one-to-one correspondence between the two puppets or the more intuitively plausible exchange. Children who were categorized at level 2 were those who fluctuated between one exchange rate and the other and who could not decide which was the appropriate one. Children were categorized at level 3 who argued that the exchange rate of the country of the transaction should be the appropriate exchange. The results clearly supported a conceptual development from level 1 to level 3 over the age groups sampled on both types of tasks. First-grade children tended to offer mostly level 1 responses and by fifth grade, the majority of children offered level 3 responses.

An Interface Between Language Background and Children's Developing Understanding of Arbitrary Properties of Base Structure Conventions. In order to determine whether language background interacts with children's developing understanding of base structure conventions as assessed with the present paradigm, we interviewed populations of third-grade Spanish-English bilinguals and another population of Chinese-English bilinguals and monolingual control subjects. Fluency in English was assessed with the Language Assessment Battery (LAB) and comprehension checks that were built into the design of our tasks. In addition to the language background data we collected on our subjects, we also thought it plausible that there may be a particular kind of cultural experience that some of these children may have that may contribute to an understanding of base structure concepts. Specifically, children from countries other than the United States may have had some experience with different currencies and monetary exchanges and thus may be ''bicultural'' with respect to monetary systems (and exchange rates) as well as bilingual. Thus, we interviewed all of our bilingual children about their knowledge of currency systems other than that used in the United States. Finally, we did two kinds of analyses of group differences, one using scores on a computational skills subtest as a covariate, the other without a covariate. Unlike the study on the arbitrary properties of number words, we did not find language background effects. However, so far we have too few subjects to partition our bilingual samples into ''bicultural-bilingual'' and ''bilingual only.'' We do see a positive trend in our bicultural group. It looks as though we may eventually find group differences such that the ''bicultural-bilingual'' children may show a significant advance over the bilingual only children, who, in turn, may show an advance over monolingual children.

Summary of Research on Intrinsic Effects

I have argued that to understand the influence of language background on mathematics achievement, one must first have a normative model of the development of mathematical cognition and a formulation of how language background may interact with children's developing cognitions. In my own work, I have shown that there is a developmental shift in children's cognitions of conventional aspects of number and that growing up bilingual does influence some aspects of children's cognitions about the arbitrary property of numerical conventions.

AN EXAMINATION OF EXTRINSIC FACTORS IN TWO CULTURAL SETTINGS

In the remaining few pages, I shift the discussion from intrinsic to extrinsic factors. The concern is with the varied ways that aspects of culture may influence and be incorporated in children's developing mathematical cognitions. I focus on two cultural contexts, each of which provides some clear illustrations that bear on this issue.

Hatano's Studies on Japanese Children's Mathematical Cognition

Japanese culture supports and places a high value on children's mathematical understandings. Such cultural support interfaces with children's developing mathematical conceptions in a number of ways as Hatano (1982) argues in his review of Japanese children's mathematics education. Hatano cites studies indicating that Japanese children prefer math to all other academic subjects. Moreover, a significant percentage of Japanese families pay to send their children to private abacus schools. (More than 2 million children every year take qualifying examinations for advanced abacus training.) Hatano notes that training in the abacus has a direct bearing on Japanese children's achievement: Hatano cites studies of his own and others that abacus training transfers to other domains of arithmetical competence like paper and pencil calculations.

Hatano also discusses linguistic factors that may facilitate the development of arithmetical competence in Japanese children. First, unlike numeration in English, Japanese number words are regular and systematic. Not only is the "ten to thirteen" sequence expressed in regular form in Japanese, the same regularity occurs for higher numbers. Thus, "three hundred and thirteen" is literally "three hundred-ten-three" in Japanese. Moreover, the regularity of the language system supports "kuku," a rhyming system each phrase of which expresses one multiplication fact of all single-digit multiplications (1×1, 1×2, . . . 2×1, . . . , 9×9). Hatano notes that all children learn this system by Grade 2.

Supports that the Japanese culture provides with respect to the value placed on mathematics, the Japanese children's cultural practices relevant to mathematics (e.g., the abacus, kuku), and the linguistic regularity of Japanese numeration each constitute a factor intrinsic to language background that is a probable source of Japanese children's mathematical competence. How any of these factors actually affects children's developing competencies are empirical questions. Research addressing these questions would help to illuminate interactive relations between factors associated with language background and mathematics achievement.

Studies on Oksapmin Children's Mathematical Cognition

Some of my own research conducted in remote "bush" schools in the Oksapmin area of Papua New Guinea provides another illustration of the way in which out-of-school experience specific to a particular cultural setting may influence what goes on in mathematics lessons in a particular school (Saxe, 1985).

The Oksapmin community is located in a remote area of the West Sepik Province. First contact with the Oksapmin by the West was in the late 1930s and it was not until the 1950s that there was sustained contact by missionaries. The first bush school was established in 1972.

The number system that Oksapmin use differs radically from the Western one. The system consists of 27 body parts and has no base structure. To count as Oksapmin do, one begins with the thumb on one hand and recites body part names as one counts around the upper periphery of the body. Children learn aspects of the system at a fairly young age. In traditional life, the body part system is used for such functions as producing a count of a set of valuables, measuring string bags (a common cultural artifact) in the course of their production, and as a means of indicating ordinal positions between two events or locations (e.g., villages on a path). There is no context in traditional life in which Oksapmin use the system for arithmetical computation. However, in areas of Oksapmin life that are being influenced by Western contact, Oksapmin are increasingly involved with computational problems.

The bush school is one context in which Oksapmin children and adolescents are presented with arithmetical problems as a part of mathematics lessons. Upon observing children in Grades 2, 4, and 6 taking an arithmetic test, it was clear that Oksapmin children were trying to use their body system to solve arithmetical problems. (Many were counting around their bodies during the test.)

In order to find out how children were using the body system to solve arithmetical problems and how the school context was influencing its use, I interviewed both children who attended school (Grades 2, 4, and 6) and children who did not attend school (matched in age with the school subjects in Grade 6). In the interview, I presented children with computational problems of different difficulty levels and observed whether, and the way in which, children used the body system to help solve the problems.

The results were clear-cut: Young children tended to use the system, but typically used an inadequate procedure to obtain a solution. For instance, to add $3 + 5$, a child might begin with the middle finger (3) and then enumerate the ring finger (4), little finger (5), wrist (6), and so on. The child has no method of keeping track of when he has counted five additional body parts and thus does not know when to stop his enumeration. In contrast, children at higher grade levels had invented an array of procedures for keeping track of body parts in computational problems. For instance, to add 3 + 5, a child might again begin with the middle finger (3). This child, however, would not call the ring finger by its actual name as the younger child did. Rather, he would call it the thumb (1), thus beginning an iteration of one to five. Thus, when the child called the inner elbow (8) the little finger (5), he would know that he had reached his answer, 8. This procedure was extended by many of the Grade 6 children to arithmetical problems that exceeded the limits of the body part system. For instance, when presented with a problem like 34 + 12, the children in Grade 6 would make use of their ability to count in both English and Oksapmin. Thus, children would say that they have "34." They would then count the thumb (1) as 35, the index finger (2) as 36, and so on, until they reached the ear (12), which they would count as 46, the correct answer. In contrast to the schooled Grade 6 children, their unschooled counterparts used virtually only inadequate body part strategies for all varieties of computational problems.

Oksapmin children's elaboration of computational procedures with their body system illustrates the way in which children are beginning to both use and adapt knowledge that is a part of their "home" culture to that culture presented in school.

SUMMARY AND GENERAL REMARKS

In this chapter, I have focused on two themes concerning the relation between language background and mathematics achievement. The first and major one concerns research approaches used to study the intrinsic effects of growing up bilingual on children's developing mathematical conceptions. I have argued for the need to embed this research question in the more general question of how mathematical concepts develop and then to construct plausible hypotheses about how language may interface with this general developmental process. Given this general approach, I have pointed to some of the methodological problems endemic to studies that address the relation between language background and mathematics achievement, including problems associated with the ex post facto design, the problem of controlling for intelligence in comparisons across population groups, and the problem with suitable operational definitions of language background. I have also offered some of my own attempts to weave through these issues in a developing program of research.

The second theme departs from the focus on intrinsic effects of language background. I have pointed out that "culture" constitutes a complex of intertwined factors—one of which is language background. Studying cultural supports for mathematics development and how children utilize different cultural backgrounds in coping with school mathematics curricula can offer insights about the sources of language minority children's successes and failures in the mathematics classroom.

REFERENCES

Ben-Zeev, S. (1977). The influence of bilingualism on cognitive strategy and cognitive development. *Child Development, 48,* 1009–1018.

Gelman, R., & Gallistel, C. R. (1978). *The child's understanding of number.* Cambridge, MA: Harvard University Press.

Hatano, G. (1982). Learning to add and subtract: A Japanese perspective. In T. P. Carpenter, J. M. Moser, & T. A. Romberg (Eds.), *Addition and subtraction: A cognitive perspective.* Hillsdale, NJ: Lawrence Erlbaum Associates.

Ianco-Worrall, D. A. (1972). Bilingualism and cognitive development. *Child Development, 43,* 1390–1400.

Piaget, J. (1929). *The child's conception of the world.* London: Routledge & Kegan Paul.

Saxe, G. B. (1985). Effects of schooling on arithmetical understandings: Studies with Oksapmin children in Papua New Guinea. *Journal of Educational Psychology, 77,* 503–13.

Saxe, G. B., Sadeghpour, M., & Sicilian, S. (1986). *Children's understanding of numerical conventions.* Unpublished manuscript, UCLA.

Vygotsky, L. S. (1962). *Thought and language.* Cambridge, MA: MIT Press.

Chapter 4

Intention and Convention
in Mathematics Instruction:
Reflections on the Learning of Deaf Students

Joan B. Stone

University of Rochester and Rochester Institute of Technology

Those of us who teach mathematics have always known that it is easier to help students develop computational ability than it is to help them develop the ability to apply basic arithmetic and algebraic operations and concepts appropriately. One way to attempt to explain this discrepancy is to consider the relationship between those intentions that we implicitly or explicitly express in the language of instruction and the content and conventions of mathematics. This relationship becomes even more complex when the everyday language of our students is not drawn from the same source as the language of instruction. It is a fundamental assumption of this chapter that language plays a role in the development of a student's ability to appropriately apply mathematical operations and concepts, and that the language of instruction may draw from the student's everyday language, but it is not the same as that language.

UNDERSTANDING MEANING

In order to understand how the language of mathematics instruction varies from everyday language, we need to think about what we mean when we express something that has mathematical content. Strawson (1971) distinguishes between two theoretical positions on the concept of meaning. He refers to these as the "communication-intention" view and the "formal semantics" view. In a discussion of the distinction made by Strawson, Feldman (1977) states, "the essential insight of the communication-intention theorists was that meaning is a property not of sentences but, rather, of their use" (p. 283). The view of the formal semanticists is that "language is used to describe states of affairs; the meaning of a description is exhausted by its truth-value . . . ; truth-value is an inherent property of sentences, not a property of the use of sentences . . ." (p. 283). The position of the communication-intention theorists has been elaborated by H. P. Grice (1971) in an article entitled, "Meaning," by J. L. Austin (1971) in his frequently quoted article, "How To Do Things With Words," and by John Searle (1971) who draws on the work of both Grice and Austin in his article entitled, "What Is a Speech Act?" It may seem ironic to claim that mathematical meaning is understood more clearly from the position of the communication-intention theorists (i.e., Grice, Austin, Searle) than it is from the position of the formal semanticists because it is the latter who have made the most use of mathematical concepts to describe their own work, e.g., Chomsky's transformational grammar. However, we have spent the last 20 years drawing on the formal structure of mathematics as a large part of the language of instruction, and it is not at all clear that the result has increased student understanding of the meaning of mathematics. The familiar and much debated "new math" was an attempt to make mathematics meaningful through the explication of its formal structure. Thus, the meaning of an operation like addition was to be found in formal properties such as $a + b = b + a$ and $a + (b + c) = (a + b) + c$. This approach to mathematics instruction falls into the same trap as that which snagged many early critics of Piaget's work. The formal properties of a system that can be observed from outside are not understood on a conscious level nor are they particularly useful to the users of the system. Young children use subject-verb-object constructions without a conscious knowledge of the definition of a verb, and they can use the commutative property of addition without the ability to say what it is they are doing. Adolescents think hypothetically without knowledge of symbolic logic. The formal truth-value of a mathematical statement is not what is meaningful to the naive learner.[1]

Contemporary views on the nature of language distinguish among its pragmatic, semantic, syntactic, and phonological aspects. The claim that mathematical meaning is understood more clearly from the position of the communication-intention theorists results in a focus primarily on the pragmatic aspect of language. It is certainly possible to consider questions about the language of mathematics instruction from the perspectives of semantics, syntax, and phonology, but such questions are beyond the scope of this chapter.

---

[1]My colleague, Professor Madeleine Grumet suggests that the forms of the disciplines are not necessarily the organizing principles in their production. She points out that research on writing shows that the structure of a text does not coincide with the process of its production.

THE ROLE OF CONTEXT

The pragmatic aspect of language refers to the role of context in understanding the content and form of language. According to Austin (1971) "The occasion of an utterance matters seriously, and . . . the words used are to some extent to be "explained" by the "context" in which they are designed to be or have actually been spoken in a linguistic interchange" (p. 563).

A fundamental concept included in the pragmatic aspect of language is that speech is an action for which the speaker has an underlying communication intent (Kretschmer & Kretschmer, 1978). It is the context of the utterance that helps the hearer determine what the underlying intent was. We are all familiar with speech acts such as "Would you like to come to the board?" which can mean either "I am asking if it would please you to come to the board" or "I am telling you to go to the board," depending on the context. This is a fairly obvious example that is quite easily understood even by students with language problems. The two possible underlying intentions, to ask something or to tell something, are easily understood even though they are not explicitly stated. Facial expressions, gestures, tone of voice, all give the hearer clues as to the teacher's intended meaning in this case. Intended meaning is much less clear when the language is not everyday language, but the language of mathematics instruction. The context of utterances in the language of mathematics instruction needs to be examined. Davis and Hersh (1981) give the following description of this context: "In an ordinary mathematics class, the program is fairly clear cut. We have problems to solve, or a method of calculation to explain, or a theorem to prove. The main work will be done in writing, usually on the blackboard" (p. 3) and "As to spoken words, either from the class or from the teacher, they were important insofar as they helped to communicate the import of what was written" (p. 3).

The claim of Davis and Hersh's that mathematics is primarily communicated through inscription rather than speech may not be entirely accurate, at least not at the more elementary levels of mathematics instruction. It seems that teachers do say quite a bit to students in terms of rules and procedures, which are then illustrated by specific written examples. Students are infrequently left to derive the rules and procedures for action from a set of examples, although in the days of "discovery learning" this did occur to some extent. However, even in those days, many teachers were uncomfortable with a process of communication that left them only asking questions. Learning theorists, such as Gagné (1966, 1970, p. 314; Gagné & Brown, 1961), attempted to reduce this discomfort and make the process more efficient by "guiding discovery" in such a way that the teacher's questions were determined by the structure of the task to be accomplished by the students. For many teachers, guiding discovery distorted communication between them and their students even more.

SUCCESSFUL COMMUNICATION

Given the context of the mathematics classroom, what would constitute successful communication between teacher and students? At the risk of oversimplifying or blurring philosophical distinctions, an answer to this question is attempted through a synthesis of the ideas of the communication-intention theorists, namely Grice, Austin, and Searle. We can say that any utterance has two parts. The first part, which may be either implicit or explicit, expresses the intentions of the speaker to produce a particular effect on the listener. This part can be understood by the listener only in reference to the context of the utterance. The second part of the utterance is the explicitly stated content. This content is understood by the listener through reference to lexical and syntactical conventions that associate it with the intended effect. Successful communication occurs when the listener recognizes the intention of the speaker, understands the explicit content of the utterance, and responds with the intended effect. An example of this would be the teacher's question, "Would you like to go to the board?" If by this the teacher meant "I am telling you to go to the board" and the student understood the intent and content of the utterance, and if the student actually went to the board, then this was a successful communication. An unsuccessful communication would have occurred if the student understood the teacher's utterance as a question and simply answered yes.

This analysis of successful communication helps to locate potential sources of difficulty in mathematics instruction. If students and teachers are from different cultural backgrounds, context may be understood differently and, as a result, speakers' intentions may be unclear or misunderstood completely. Also, the conventions of mathematics may diverge sufficiently from the conventions of the student's everyday language so that even if the teacher's intentions were explicitly stated, the student may not understand how to produce the intended effect.

We have perhaps focused on convention within mathematics in our research on mathematics learning at the expense of intention. If we consider the operation of addition as an example, we can see that there is a big difference between knowing how to add and knowing when to add. In the case of knowing how to add, convention and intention coincide. The teacher says "Three and five is eight." (One needs to think what the verb is in this sentence. In ordinary language, the plural form is required; however, the mathematical sentence calls for the singular because the unspoken subject is "The sum of . . . .") For the student, the problem is simple. Knowing the convention "Three and five is eight" and producing the intended effect when the teacher says "Three and five is _____," are the same thing. The lexical and syntactical character of the teacher's utterance, "Three and five is _____," conventionally associates it with the intended effect. When convention and intention coincide, one would expect that students whose everyday language is different from the language of instruction would have fewer difficulties than they would if this were not the case.

DEAF STUDENTS' LANGUAGE AND THE LANGUAGE OF MATHEMATICS

Assuming that tests of arithmetic computational ability are representative of situations where convention and intention coincide, it is useful to look at the performance of deaf students on such tests and to compare that performance with their performance on tests of arithmetic applications and concepts where there is more likely to be a discrepancy between everyday language and the conventions of mathematics. The majority of people in the United States today who are deaf were either born deaf or became deaf before they acquired language. As a result, the English language reading and writing ability of deaf students is quite low. The average deaf student leaves high school reading English at a third-grade level or lower. According to Moores (1978), demographic studies of the academic achievement of deaf students as measured by the Stanford Achievement Test indicate that "without exception, higher grade equivalent scores are earned for Arithmetic Computation, which requires computational skill but little knowledge of Standard English, than for Arithmetic Applications or Arithmetic Concepts, subtests that place more emphasis on reading" (p. 225). The tendency for deaf students to know better how to perform a mathematical operation than when to perform it is verified by studies conducted at the National Technical Institute for the Deaf at Rochester Institute of Technology. Daniele, Sachs, and Bonadio (1986) report that

> Performance levels of entering students at both the National Technical Institute for the Deaf and Gallaudet College indicate a serious need for remedial work in basic arithmetic, introductory algebra, and rudimentary thinking skills. Far too many students view mathematics as being only computational in nature. (p. 2)

The difference between computational ability and appropriate application of basic arithmetic and algebraic operations and concepts appears to be significant in those students for whom the language of instruction is different from their everyday language. It might be argued that the difference between the language of mathematics instruction and the student's everyday language creates a problem even when both are drawing from a common source such as English. It seems obvious that this problem is compounded when the language of instruction draws from English and the student's everyday language is different from English, as it is for the child of Spanish-speaking parents for whom the everyday language at home and in the community is Spanish or the deaf children of deaf parents for whom everyday language is American Sign Language (ASL). In the case of some deaf students who have grown up outside a deaf community, the problem is further complicated by an everyday language consisting of a few easily understood English words and some idiosyncratic signs that are not understood beyond the confines of the student's own family.

It would be a distortion to say that mathematics consists of nothing but computation even at the most elementary level. Knowing how to add is trivial, if one does not know when to add. Often students will look for extraneous cues to help them decide if a particular problem can be solved by addition, e.g., "This is page 234, so it must be addition." Other times they will look for function words such as *and* or *of* to tell them what to do. Here English language convention may diverge from mathematical convention, and both may be different from the intention of the speaker (writer) of the problem. Davis and Hersh (1981) discuss a set of word problems that "ostensibly call for addition" (p. 71).

> Problem 1. One can of tuna fish costs $1.05. How much do two
> cans of tuna fish cost?

The solution appears simple. Either we add $1.05 and $1.05, or we multiply $1.05 by two. Mathematically, the solution is $2.10. Davis and Hersh point out, however, that markets do not always price goods in this way. Frequently, the price of one can is $1.05, but the price of two is $2.00. "Discounts are so widespread that we all understand the inadequacies of addition in this context" (p. 72). It is not clear that everyone would understand the inadequacy of addition in this problem, but the possibility for confusion exists on several levels. First of all, ordinary language convention would suggest that the problem calls for something other than simple addition. Davis and Hersh are correct in their analysis of the prevalence of discounts. Knowing that one item costs a particular amount does not always help one determine what more than one of those items would cost. Second, mathematical convention is not clearly determined in this problem. Multiplication works as well as addition. In fact the use of the word *of* may indicate multiplication to many readers of the problem. All of this leaves the student dependent on the context of the problem to determine the intention of the writer. Typically, the context of such problems consists of a list of similar problems where the intended effect will be relatively clear in perhaps one or two problems. Thus in the example above, if the problem appears on page 234, then the teacher (writer) must have intended that the student should add. This is hardly the kind of meaningful context that will lead to an understanding of when to add.

ASSOCIATING INTENTION AND CONVENTION

This analysis illustrates the problem of the divergence of intention from either mathematical convention or ordinary language convention for any student. The problem, understood in this way, is obviously exacerbated when the student's ordinary language is different from English. The language of instruction includes the intersection of ordinary language conventions and mathematical conventions. If students do not know the conventions of ordinary language or if they have a limited or distorted understanding of these, then what the teacher intends when she or he uses the language of instruction will be partially understood at best. The problem for the teacher is to establish the lexical and syntactical character of instructional utterances in such a way that these utterances will become conventionally associated with the intended effect. Those of us who teach mathematics to deaf students at the Rochester Institute of Technology, National Technical Institute for the Deaf, have more or less consciously been trying to do this for years. Within our individual classrooms, we are relatively successful in establishing linguistic and mathematical conventions that are shared by the majority of our students.

The problem occurs, however, when deaf students are asked to "apply" the mathematics we have taught them in technical areas such as architecture and civil engineering. The conventions we have established within the mathematics classroom to coincide with our intentions as teachers of mathematics do not function in the same way, if they exist at all, in engineering classrooms. The students perceive that they are in a completely different situation and that their mathematical abilities are not obviously applicable. For example, in mathematics we teach the Pythagorean theorem in several contexts. We teach the formula in algebra so that students can use it to practice working with irrational numbers. We teach the proof of the theorem in geometry as part of a sequence of theorems about similar triangles. We also use the Pythagorean theorem again in trigonometry to establish identities. These students have certainly encountered the theorem in a variety of different situations and, if all the concept formation literature is correct, we should expect that the concept of the Pythagorean theorem would "transfer" from these various mathematical situations to the engineering classroom. In fact, it does not. Students who are capable of fairly sophisticated mathematical reasoning do not "see" the practical application of something as basic as the Pythagorean theorem in simple surveying problems. Our language of instruction is not meaningful to our students outside of the context of our mathematics classrooms.

Lists of objectives, which describe what students will do, do not clarify the intentions of the teacher by themselves. It is difficult for the students to recognize what was intended when the teacher is not sure what he or she is doing by saying a particular thing in a classroom. If meaningful communication is to occur in the domain of mathematics, then the first step is for the speaker to have a definite intention by his/her utterance. To go back to the Pythagorean theorem example, we can see it had an "oh, by the way" quality about it. In each case, our intention was to communicate something other than the theorem itself, by using the theorem. Students were expected to understand not only the theorem, but its underlying nuances as well, although it was never explicitly stated that they do so. When we look at the situation this way, it is no wonder students did not recognize the application of the theorem in a different context. It never was our intention that they should do so. Asking students to extract unintended meaning is an unpredictable business.

What is it that we do in mathematics classrooms with students who have language problems? It is possible that in the face of difficult communication, we have a tendency to eliminate as much conventional language as possible. We rely on diagrams and gestures that may have little or no meaning outside of our classrooms. Linguists who study American Sign Language have warned those of us who teach deaf students to limit the number of situation specific signs we create for use only in our classrooms. Their concern appears well-founded. If Austin (1971) is correct that "the occasion for an utterance matters seriously" (p. 563), then we need to take care that those occasions are as rich as possible. Elimination of written and spoken language hardly enriches the context of instruction for students who have language problems, although in the short run this may seem to ease the task of the teacher. It is human nature to try to make a complex situation more simple, but finally the simple reduction of language in situations where communication is difficult as it may be with deaf students does not result in greater learning. It is important to remember that a great deal of incidental learning goes on for students for whom the language of the classroom and everyday language are the same. To increase the opportunities for incidental or unplanned learning for deaf students and other students for whom the classroom language is a second language, greater exposure to the language of the classroom and the language of mathematics is necessary. Rather than eliminate, we need to elaborate language for these students. More, not less, communication is required for intention to be recognized, for convention to be broadly defined, and for understanding to occur.

REFERENCES

Austin, J. L. (1971). How to do things with words. In J. F. Rosenberg & C. Travis (Eds.), *Readings in the philosophy of language* (pp. 560–579). Englewood Cliffs, NJ: Prentice-Hall.

Daniele, V., Sach, M., & Bonadio, A. (1986). *NTID Outreach Project: Task force for mathematics education*. Unpublished report, Rochester Institute of Technology, New York.

Davis, P., & Hersh, R. (1981). *The mathematical experience*. Boston: Houghton Mifflin.

Feldman, C. F. (1977). Two functions of language. *Harvard Educational Review, 47*(3), 282–93.

Gagné, R. M. (1966). Varieties of learning and the concept of discovery. In L. Shulman & E. Keislar (Eds.), *Learning by discovery: A critical appraisal* (pp. 146–147). Chicago: Rand McNally.

Gagné, R. M. (1970). *The conditions of learning*. New York: Holt, Rinehart & Winston.

Gagné, R. M., & Brown, L. T. (1961). Some factors in the programming of conceptual learning. *Journal of Experimental Psychology, 62*, 313–21.

Grice, H. P. (1971). Meaning. In J. F. Rosenberg & C. Travis (Eds.), *Readings in the philosophy of language* (pp. 436–444). Englewood Cliffs, NJ: Prentice-Hall.

Habermas, J. (1979). *Communication and the evolution of society.* Boston: Beacon Press.

Kretschmer, R., & Kretschmer, L. (1978). *Language development and intervention with the hearing impaired.* Baltimore: University Park Press.

Moores, D. (1978). *Educating the deaf: Psychology, principles, and practices.* Boston: Houghton Mifflin.

Ricoeur, P. (1976). *Interpretation theory: Discourse and the surplus of meaning.* Fort Worth: The Texas Christian University Press.

Searle, J. R. (1971). What is a speech act? In J. F. Rosenberg & C. Travis (Eds.), *Readings in the philosophy of language* (pp. 614–628). Englewood Cliffs, NJ: Prentice-Hall.

Strawson, P. F. (1971). Meaning and truth. In P. F. Strawson (Ed.), *Logicolinguistic papers* (pp. 170–189). London: Methuen.

Chapter 5

Why Should Developmental Psychologists Be Interested in Studying the Acquisition of Arithmetic?

Ellin Kofsky Scholnick
University of Maryland

In one of the Sesame Street books, the Count gives up enumeration after he discovers he has made an error. This leads him to a second, important discovery. Every substitute occupation he could enter also requires counting. Quantitative thinking plays such a prominent role in a technological society that those who have difficulty acquiring and applying quantitative concepts may be barred from access to prestigious and vital technological careers. Because a significant part of the curriculum is devoted to teaching mathematics and topics that depend on mathematical thought, educational psychologists have been interested in studying the structure of mathematics, children's understanding of it, and the design of an effective instructional program tailored to children at different levels of understanding. They have also been concerned with how the cultural and linguistic background of students may affect acquisition of mathematics.

Recently developmental psychologists have also turned their attention to the field of mathematics learning. The importance of mathematics in education and daily life is not the sole factor motivating them. Cognitive developmentalists study mathematics learning because they continually debate assumptions about the nature of the human mind, the structure of knowledge, and the process of acquisition. The properties of mathematics make study of its acquisition an ideal place to address those controversies. This chapter begins with a discussion of why models of mentation and models of mathematics are so intertwined that the study of one may require study of the other. It then proceeds to delineate particular facets of the structure of arithmetic, a small part of mathematics, that enable insighis into the nature of mental development. Many issues in this section are abstract because mathematics is used as a way of examining broad theoretical questions that go beyond mathematics learning. The third section draws attention to the particular analyses to which this book is specifically addressed—the role of language and culture in the acquisition of mathematics.

THE CENTRALITY OF MATHEMATICAL THOUGHT

One portion of cognitive psychology is concerned with the process of logical and scientific reasoning and the structure of knowledge in those domains. Because mathematicians have drawn on analyses of logic to delineate their field and cognitive psychologists have relied heavily on quantitative models to describe mental processes, the two fields overlap and mathematical thinking is an important area of psychological research.

Since mathematics is the product of human cognition, it is inevitable that the study of mathematics and the study of cognition converge. The structure of mathematics has sometimes been equated with logic due to mathematicians' reliance on logical proofs to validate analyses and to the prevalence of theories of mathematics that draw on logical relations such as sets. Mathematical thinking draws on two processes (Gardner, 1983; Kline, 1980): search for consistent patterns and principles (induction), and the construction of chains of inference to support arguments (deduction). A broad spectrum of cognitive theories (e.g., Klahr, 1984b; Piaget, 1977; Schank, 1982) use those same processes to explain development. The child is assumed to acquire some knowledge and construct some rules just as the mathematician does, by induction and deduction. This has led cognitive psychologists to study mathematical thought because it is a realm that is very representative of human thought or because certain logical relations that have been the subject of much cognitive research, such as class inclusion and transitivity, also underlie mathematical thought.

Additionally, psychologists often build mathematical models of the content and acquisition of conceptual domains and use measurement scales to validate their analyses. Piaget (1965, 1971) has used arithmetic logic as a model for cognitive organization, and Norman Anderson (e.g., Anderson & Cuneo, 1978) has described the integration of information from different sources as either additive or multiplicative. Cognitive science characterizes humans as manipulators of symbol systems according to a set of delimited and well-specified rules that have mathematical properties. The assumption that children act as if they were conceptualizing or structuring information in mathematical terms leads to the investigation of the child's understanding of those terms.

For those psychologists who are interested in using mathematical models to characterize human thought, the investigation of mathematics is particularly appropriate because the language of the model and the language of the domain to be modeled are compatible (Ginsburg, 1983). As a consequence, the study of mathematics learning is also conducive to evaluation of the efficacy of formal modeling as a psychological tool. The validity of those models is debatable. Not everyone agrees that people operate within a domain by using the formal principles by which the domain is presumed to be structured. Debates continue about the extent to which individuals obey the principles of logic in acquiring and reasoning from information (e.g., Braine & Rumain, 1983; Johnson-Laird, 1983). It is unclear whether people reason from conditional premises by using a truth table, by memorizing certain reasoning rules, or simply by using their knowledge of the world. Similar debates arise about whether everyday calculations are based on an understanding of arithmetic that mirrors the mathematician's. If people do not perceive the formal structure of arithmetic as a set of interconnected principles that allow no violation or if the child does not understand the logic of arithmetic, then it makes no sense to contend that the expert's model of arithmetic is the child's model or that mathemetical logic governs how the child understands and operates with number or builds notions of time, space, and physics. Models in psychology serve as precise formulations that summarize the organization of a set of laws. They may also be thought to have psychological validity. The ease of modeling arithmetic knowledge in the same language as its content makes mathematics learning a good place to test fundamental questions about the limits of formal analyses of domains. Although this concern seems far removed from the educational enterprise, it is not. We have only to remember the "new mathematics" where the emphasis in arithmetic curricula shifted from the teaching of addition facts to the teaching of set relations and principles of reasoning.

ARITHMETIC AND DEVELOPMENTAL PSYCHOLOGY

The preceding section presents the claim that mathematics learning is central to the study of cognition. In this section, we explore several properties of arithmetic that make it particularly suitable for research into controversial issues in developmental psychology. The controversy arises from competing views of development. Logical constructivism characterizes children as problem solvers who seek to discover the general principles that organize knowledge in any content area and who arrive at those principles by induction and deduction without the aid of formal instruction (e.g., Piaget, 1977). Opposing views do not claim that children construct theories, but rather that children simply learn to use specific skills in specific situations. The child does not learn an arithmetic but many arithmetics, and there are many avenues to learning (e.g., Siegler & Richards, 1983; Siegler & Robinson, 1982; Siegler & Shrager, 1984). The nature of arithmetic is so conducive to constructing a logical theory that if the child does not behave as expected in learning arithmetic, it is unlikely that the child approaches any domain that way.

Mathematics as a Formal System

Because the formal structure of mathematics is consistent, explicit, deterministic, and delimited, study of its acquisition has advantages. The availability of explicit theories of arithmetic allows testable theories of how that domain is construed and how knowledge of it is acquired. When problems are open-ended, the many potential solutions to them impede building a model that accounts for all of them. Arithmetic problems are both delimited and rich enough to provide a small but interesting set of hypotheses that the child can entertain. There are promising models of what the child must do to count (e.g., Greeno, Riley, & Gelman, 1984), or to add or subtract (Resnick, 1983) that can account for mistakes as well as successes (Brown & Van Lehn, 1982). The existence of models for competent and less competent performance also allows inferences about what is necessary for a transition to mastery and the design of instructional programs to enable students to make the transition. Evaluation of these instructional programs can validate theories of acquisition. Hence a well-specified domain lends itself to analyses of development within it. Explicit analysis of an area also facilitates choice of what to study. For example, Peano's analysis of the centrality of ordinal principles in arithmetic initially led Gelman and Gallistel (1978) to examine how the child orders countables and counting symbols and establishes a correspondence between them. It also led Brainerd (1979) to test the centrality of ordering and transitive relations in understanding concepts of relative quantity.

Finally, a well-specified domain lends itself to the study of individual differences. This book argues that the structure of a domain, and the operations necessary to function within it, may not be the only influences on acquisition. Acquisition is also affected by the meaning systems an individual imposes on input. Theoretically, the difference between 2 and 3 is the same as that between 3 and 4, but young children often find it harder to make relative magnitude judgments of 2 and 3 because they think of those two numbers as more similar than 3 and 4 (Siegler & Robinson, 1982). The Oksapmin in Papua New Guinea represent numbers as body parts and tend to confuse two numbers that are designated by symmetrical body parts such as the left and right eye (Saxe, this volume; Saxe & Posner, 1984). The study of a well-specified domain makes it easier to detect how the individual's system of meaning influences understanding of the domain as well as to delineate the nature of the material to be learned and the process by which it is learned.

The Clarity and Reliability of Mathematical Concepts and Laws

The clarity of mathematical concepts also makes mathematics a very good field to study learning in its purest form. Many natural concepts are fuzzy because they overlap and lack clear criteria defining them (e.g., Rosch & Mervis, 1975).

The aquatic whale is actually a mammal, not a fish, though the majority of mammals are land dwellers. Thus some atypical instances of natural categories may have to be acquired by rote. In contrast, because most arithmetic rules permit no exceptions, they can be learned by induction. It should also be clear when errors occur. There is only one solution to $2 + 3 = x$, and it will always be the same one. Two different answers to the same problem indicate an error. The constancy of arithmetic might facilitate acquisition of particular facts and the development of algorithms. The predictability of outcomes might also foster the skills in planning and monitoring that are thought to be fundamental to cognitive development and the management of cognitive resources. Therefore, arithmetic is an ideal place to study learning (e.g., Siegler & Schrager, 1984), rule induction (Riley, Greeno, & Heller, 1983), and planning and monitoring (Van Lehn, 1983). There has been much effort devoted to constructing sets of geometric figures for concept learning tasks and lists of nonsense syllables to study associative processes, but mathematics provides a natural setting for research on concept and rule learning. The study of mathematics learning also gives rise to a provocative question: Do children ever become aware of the reliability of definition of concepts and use that awareness to simplify the task of learning?

Mathematics Expresses Principles of General Applicability

The laws of arithmetic apply to each number because each number is produced by iteration. Consequently arithmetic is a good field in which to examine the degree to which individuals can appreciate the generality of principles. Developmental and cognitive psychologists often debate the extent to which thinking should be characterized in terms of general cognitive structures or in terms of packages of specific skills assembled to meet specific task demands. The issue is fundamental. Without some generality in behavior, each situation would be novel, yet if we were only generalists, we might fail to detect crucial nuances that distinguish one situation from another. In order to determine whether the child appreciates the general applicability of cognitive structures, there must be agreement that analogies exist across domains, but there are often disagreements. Is judging the perceptual field of someone standing opposite from you similar to understanding what must be communicated to students about a homework assignment? It depends on whether you think that taking the spatial perspective of another is the same as adopting the person's conceptual perspective. Unlike the study of egocentrism, mathematical connections are often straightforward. If the child does not generalize mathematical knowledge from one set of numbers to another, in a domain which is so connected, then it is unrealistic to claim that development is governed by general cognitive structures. Arithmetic learning should allow examination of the conditions under which children can apply past learning to new situations.

When generality breaks down, as may be the case in early number learning

(Siegler & Robinson, 1982), the opportunity arises to explore what limits application of principles. Two kinds of explanations can be examined. Sometimes, generalization ought to occur because the same operation is used in different amounts. Borrowing from zero requires adjustment of more adjacent columns than borrowing from other numbers (Resnick, 1983). Multiplication of 22×55 differs from multiplication of 22×555 only in the number of digits to be manipulated. Children may understand subtraction and multiplication but they may fail the task that is more demanding because of limitations on processing capacity. The failures are particularly likely to occur at the initial stages of learning when children are still devoting so much effort to retrieving number facts that they cannot handle many arithmetic operations (e.g., Case, 1985; Wilkinson, 1984).

There are also qualitative explanations for the failure of generalization reflecting conceptual misunderstanding. My son who is an expert in subtracting $35 - 19$ was puzzled when he had to subtract $3\frac{1}{2} - 1\frac{9}{10}$. His subtraction algorithm consisted of taking 1 from the column to the left and adding to the column on the right but he did not see its relevance to fractions. He did not realize that in each case, he was dealing with conversion of units from one base to another. In the first example, you are adding 10 so you have $20 + 15$ from which you subtract $10 + 9$. In fractions, you have $2 + 1\frac{1}{2}$ (that is, $2 + \frac{15}{10}$) from which you subtract $1 + \frac{9}{10}$. His conception of borrowing failed to enable him to see the analogies. The development of expertise has sometimes been characterized as creating abstract concepts that permit analogies between seemingly disparate problems (e.g., Chi, Glaser, & Rees, 1982).

Arithmetic enables examination of the ability to make two sorts of generalizations: (1) use of the same operations when there are more steps or a larger quantity and (b) use of the same operation in fields that appear similar only if the meaning of the task is redefined. In a field where there are highly connected concepts and highly generalizable laws, we can examine the task conditions and stage of learning at which connections might be forged and become conscious (Gelman & Baillargeon, 1983). Arithmetic learning also enables the study of the sequencing of qualitative and quantitative extensions. In summary, whether one wishes to test if cognition is a single entity or examine the conditions under which already discovered rules can be used elsewhere, arithmetic learning may provide important data.

Mathematics as a Connected System

Mathematics is also integrated through tight interconnections among operations, among numbers, and between numbers and operations. Those multiple links permit the child to learn arithmetic through many routes. One set of connections arises from the nature of number itself. Because numbers are composed of other numbers, the order of combining parts is irrelevant to the outcome. The distributive and commutative properties of arithmetic produce multiple ways to do arithmetic problems. Addition of a column of numbers can be done in order or by first finding combinations that sum to 10 before proceeding. As long as the addends are exhausted, the order of additions is irrelevant. When there are several ways to solve a problem, children who are taught one strategy frequently derive other strategies by taking into account the properties of arithmetic (e.g., Ginsburg, 1977; Groen & Resnick, 1977). It then becomes possible to study the level of knowledge children must attain before inventing new strategies and the basis for their inventions. Additionally, the multiplicity of strategies for problem solution forces the child to choose among strategies. Siegler and Shrager (1984) have examined how children tailor particular strategies to particular arithmetic problems. Often children first use the simplest strategy and then fall back on more complicated ones when the simpler ones fail. Thus examination of children's performance yields vital information about their evaluation of the utility of strategies and the effectiveness of their own efforts. Those evaluations may change during acquisition.

Another set of mathematical connections arises because arithmetic consists of a small set of operations that are put into diverse combinations and applied to an infinite variety of numbers. Long division incorporates division, multiplication, and subtraction. In this respect, learning arithmetic resembles learning to type or drive. In each case, after acquiring a simple skill, that skill is embedded into a larger unit. As each part becomes well learned, it requires less attention and it is coordinated more smoothly with the rest of the routines. Research on arithmetic learning can provide data on the conditions under which children can assemble a module and embed it into another routine.

Arithmetic has been described as a collection of facts, as an organization of concepts, and a set of procedures that, like components of an aerobic routine, are assembled into smoothly functioning units. Whatever the view of acquisition, there seems to be a facet of arithmetic to which that theory may be applied. The excitement of research into arithmetic is that those diverse views can be integrated because arithmetic skill differs from motor skills, and the structure of arithmetic is not exactly like some other cognitive domains. In an exercise routine, each step is integrated into a whole and loses its identity as a separate component. In performing long division, which involves successive divisions, multiplications, and subtractions, each operation may not lose its separate identity but it gains in meaning. The routine provides information about the nature of set partitions and the inverse relation between multiplication and division. The movement of an operation into a new context can enrich understanding of that operation. Arithmetic learning may involve a continual reworking of knowledge. The reworking is due to a third kind of connection, the close link between the numbers that comprise arithmetic and the operations on these numbers. The mathematical system makes learning to calculate different from learning to drive. The procedures that govern addition are not just strategies. They flow from the properties of number. Two is the result of addition of 1 + 1. Because it is the result of more additions than 1, 2 is greater than 1. Arithmetic unites conceptual and procedural knowledge. In the study of memory, there is a division between strategies employed to enable retention and the content retained. Some of the most effective strategies make something meaningful or distinctive, but they do not define meaning. Mathematics has no such artificial distinction, so the study of arithmetic learning permits scrutinizing how the acquisition of a procedure may at some time lead to enrichment of a concept or fact, and how the acquisition of understanding of new conceptual properties gives rise to new procedures. That connection allows us to integrate in the study of arithmetic mastery diverse models of learning and the reworking of knowledge that can occur because the facts and procedures learned have implications for one another.

Mathematics as a Rich System

Because each numerical concept is embedded in a system, it has many meanings. Thus the study of number development enables examining the derivation of these multiple meanings and individual differences in routes to mastery. For example, Fuson and Hall (1983) list several interpretations of the word *two*. The number *two* on a player's jersey is used nominally as an identifier or code. *Two* can have an ordinal reference to relative position. *Two* also denotes numerosity. In terms of continuous quantity, *two* does not refer to the number of things but to the number of uniform, repeatable units as in *two quarts*. *Two* also exists as a sequence word, a member of a list of words just as *re* is a member of the list of musical notes. Those meanings are discriminable, yet they are interrelated. Consequently, understanding one aspect of the meaning may help in interpreting another. For example, although a number may designate a numerosity, the understanding that one numerosity is greater than another may be facilitated by the ordinal position of the number word in the counting sequence. Similarly, when the child understands that measurement terms refer to the iteration of standard units, then the child may understand more about the process of addition (Cooper, 1984). Thus the study of arithmetic allows tracking the emergence of interconnections and meanings in a complex and rich semantic field.

Like many semantic concepts, the development of number meanings has been explained in diverse ways. In the enrichment approach, the child begins with a single means of understanding, e.g., subitizing (Klahr & Wallace, 1976) or counting (Gelman & Gallistel, 1978), which is the elaborated in a fixed progression. Alternatively, children may endow number words initially with many meanings, some of which are redundant or irrelevant. Irrelevant ones are discarded and other meanings are discriminated according to the context of their use, and then those contexts are interrelated (e.g., Fuson & Hall, 1983). Since the enrichment and differentiation explanations have been applied to many aspects of perceptual and semantic development, the acquisition of number meanings provides another way to evaluate the validity of each position. Moreover, as in the study of the acquisition of syntax (Maratsos, 1983), the richness of number meanings may produce individual differences in the route by which those meanings are analyzed or elaborated. These individual differences have implications not just for theories of development but also for curriculum design.

Mathematics as a Symbol System

Abstraction.   Cognitive psychologists have been attracted also to the study of mathematics because its abstract, symbolic properties are common to many representational systems. Numerosity does not reside in objects but in relations among ordered sets. Numbers designate continuous and discontinuous quantities, and perceptible as well as hidden arrays. One can even assign numbers to numbers as in the solutions of the missing addend problem $(4 + \underline{\hspace{1cm}} = 6)$ through counting "5 is 1 and 6 is 2." Numbers can be converted into other base systems without changing their referents. Therefore the study of arithmetic learning enables the researcher to track the development and use of abstractions and to explore the environmental factors that influence acquisition and use (e.g., Davydov, 1982; von Glasersfeld, Steffe, & Richards, 1983).

Infants and preschoolers seem to represent numerosity in terms of perceptual patterns or counting operations. Von Glasersfeld and Richards (1983) have described the conversion of this practical, tangible, informally organized, perceptually based counting system into a symbolic one. The child's idea of what can be counted changes from a visible object to a pointable location, then to a verbalizable symbol, and finally to an abstract symbol itself capable of being decomposed or joined, or being symbolized further. Often, the manipulation of mathematical concepts in long chains of reasoning far removed from the data is especially difficult for students. It is interesting to speculate about whether the concept of automaticity applied to motor skills applies to the manipulation of abstract concepts.

Symbolization.   Arithmetic is a three-tiered system, consisting of the quantity to which the child refers, and its verbal and written symbols. As with many symbol systems, some of the links among written and spoken numbers and their referents are arbitrary. The written number 9 contains no clue that it refers to a larger amount than 8. The word *nine* bears no phonological resemblance to the written numeral 9. For some terms, such as *nineteen,* the order of the spoken digits and the written ones (19) are inverted. Yet, there is a syntax that underlies the writing of integers. The rightmost digit designates multiples of 1 and each successive digit to the left indicates the number of units at a higher power of 10. Learning the notational system requires awareness of the base-10 system and of the role of zeros as "place holders." The zero in 609 means there are no 10s.

Every notational system requires learning to symbolize some referent and to understand the syntactic rules governing the order of combining symbols. Consequently there are three important research issues that can be addressed. First, there may be a common core of demands in the acquisition of notational systems such as learning how a referent is notated and how to combine symbols. Acquiring any form of literacy may also foster a common set of skills, such as critical analysis of arguments and procedures. Second, there may also be unique problems in learning specific systems of notation and unique outcomes due to their distinctiveness (Gardner, 1983). Perhaps the emphasis on componential analysis and linear ordering in arithmetic fosters an analytic, orderly style of problem solving. Sometimes learning one kind of notation may interfere with using another. That interference may account for the difficulties in translating word problems into algebraic equations (e.g., Riley et al., 1983). Third, mastering some number notation systems may be easier than others. Both Roman and Chinese numerals are more iconic than Arabic numbers. The number of strokes in the written symbol increases from 1 to 4 corresponding to increases in the quantity. English uses one verbal system for counting quantities but the Japanese have different counting suffixes depending on the shape of what is being counted.

Thus the study of arithmetic thinking provides a chance to examine the acquisition of a notational system, a particular form of representation. The opportunity exists to determine whether acquiring numerical literacy resembles the acquisition of other forms of literacy or is unique; whether mastery of one notational system facilitates or interferes with mastery of others, and whether there is a difference between mastery of different number notational systems. Earlier portions of this chapter posed the question of whether the child behaves as if the laws of arithmetic hold across various numbers and whether the child uses the understanding of order and inclusion relations in arithmetic with the same skill when dealing with temporal or semantic relations. That discussion of skill generality is broadened by examining whether literacy produces common or specific outcomes and draws on specific or general skills.

Number Pervades the Life Span

The preceding sections emphasize the abstract properties of arithmetic that lend themselves to the study of cognitive development. However, properties of the mathematics learner also contribute to interest in the field. Developmental psychology is often a study of origin, a search for the earliest inklings of a behavior or concept in order to trace its development and choose the point at which instruction can begin. It is often hard to discern in infant behavior intellectual content similar to adult performance on academic tasks (e.g., Bornstein & Sigman, 1986). A notable exception is infant discrimination of number (Antell & Keating, 1983; Cooper, 1984; Strauss & Curtis, 1984). In several studies, infants were shown a given number of objects. From trial to trial the objects and their arrangement varied, but the quantity did not. After habituation, the baby saw either the same number of objects or a different quantity. Even within the first half year of life, babies discriminate between small quantities such as two and three and recover from habituation when a quantity different from the one to which the infants were habituated is shown. Somewhat older infants who hear two beats are more attentive to a visual display with two objects than displays with other quantities (Starkey, Spelke, & Gelman, 1982). No one would claim infants are capable of balancing a checkbook or even knowing that 2 is less than 3. The ability to differentiate quantities does not even extend to distinguishing 4 from 6. However, signs of some form of number appreciation early in development raise the possibility of stable individual differences in infants that might predict later developmental trajectories. It then becomes possible to examine the causes of the continuity and the means by which knowledge is transformed. One potential direction of prediction is that babies who rapidly discern the commonality among various sets of two despite the diverse makeup of the sets and who react strongly when the quantity in a set is changed may, at a later age, be among the first who are able to discriminate larger numbers from one another and judge the relative magnitude of sets. The challenge is to determine the extent to which early individual differences in primitive perceptual discriminations of number are predictive of the ease with which the child grasps more complex and abstract number relations and to determine how far the chain of predictability extends. This is an important issue because such predictions enable detection of children with mathematical talents or disabilities. Additionally, one proof that arithmetic is an integrated body of knowledge is the accurate prediction of performance from one level of conceptual difficulty to the next. A second possibility is that predictability in development or continuity is not due to mathematical ability but to some general facility in information processing. Thus the ability to attend to ever-changing sets and to discern their commonality may be more important than whether the sets differ in numerosity or whether they are categories of animals (Bornstein & Sigman, 1986). The appearance of a behavior or ability in both infants and adults enables attempts to trace and explain its developmental course and to trace the emergence and stability of individual differences.

The early appearance of quantity discrimination suggests the existence of an innate capacity for quantification (e.g., Klahr, 1984a) and supports the view that we are born with a set of preprogrammed capacities for dealing with a narrow range of contents, such as number. Development consists merely of gaining easier access to those skills, being more reflective about them, and also learning to use them more fluently and in broader cognitive contexts (e.g., Fodor, 1983; Gelman & Baillargeon, 1983). It then becomes important to evaluate the extent and nature of environmental influences on the application and transformation of mathematical knowledge.

Finally, quantity not only enters the life of infants and students, but also adults who use arithmetic in their occupations and in daily transactions. The extent to which those continual encounters with quantitative tasks change people's understanding of and practical skill in mathematics is a question that is just beginning to be explored (e.g., Scribner, 1984).

Number is Embedded in a Sociohistorical Context

The preceding discussion of arithmetic assumes that it is a fixed, already charted body of knowledge based on universally acknowledged and timeless principles. Actually, our theory of arithmetic is still evolving. Ancient number theory was imbued with religious speculation about the significance of particular numerals. The logical analysis of the principles of arithmetic began in the late 19th and early 20th century. Even today there are disagreements about the components of arithmetic logic and the utility of formal deductive systems in mathematics (Kline, 1980). Arithmetic procedures have also changed historically. Egyptians performed multiplications by adding doubles of numbers. The absence of zeros and place values in Roman numerals produced arithmetic procedures quite different from ours. It was not until the 16th century that Western Europe adopted from the Arabs place value notation, and the set of arithmetic algorithms based on them evolved. Thus the connections that we presume to exist in the mind of the contemporary arithmetic expert were absent in earlier eras.

The structure of mathematics as we know it today and the methods of calculation that we teach are indeed cultural inventions or discoveries. The uses to which we put arithmetic constantly change. History should make us both humble and curious about our analysis of the nature and origins of mathematical abilities. Historical change in the understanding and use of mathematics provides the opportunity to explore the impact of such changes on thinking. One interesting example is Saxe's research (Saxe & Posner, 1983) on how the emergence of the Stone Age Oksapmin in Papua New Guinea into a trade-oriented culture has changed the nature of their number system and presumably their conceptions of number relations. The number system that was originally represented by the position of body parts is slowly evolving into one with a base system reflecting the monetary units Oksapmin need to exchange. Because the new number system is much more abstract than the body-part one, there is a chance to examine parallels between historical and ontogenetic development by comparing number understanding among adults and children who may differ in exposure to commercial transactions. Perhaps the child at the trading post has a more sophisticated understanding than the adult Oksapmin farmer.

Modern reliance on calculators and computers may also produce changes in arithmetic understanding. Calculators deemphasize fact retrieval while calling attention to procedures for data entry and manipulation. They emphasize outcomes and, as a result, enable induction of the regularities that produce outcomes. Performing an analysis of variance with a calculator quickly makes clear that the sum of the squares of 10 + 10 is a much smaller number than 15 + 5 or 18 + 2. Computing the same results by hand is so effortful that less time is spent on comparing outcomes. Since calculators are machines, their users learn to check data manipulations to see if the machine is in working order. In contrast, although there are numerous checking procedures in arithmetic that people should employ, the routines require so much effort that people are not likely to use them. Checking the outcome of long division by multiplication is as effortful as the original, already burdensome calculation. Thus the study of mathematics learning provides opportunities for examining the interplay of technology and mathematics understanding. Presumably these data might have implications for the evolution of thought in general.

In the historical evolution of mathematics, different societies have designated numbers and taught arithmetic in different ways. We have noted the difference between the Oksapmin and Arabic counting systems, and the complex system of counting suffixes present in Japanese, but absent in the English language. The system of Roman numbers that works on multiples of 5 is very conducive to finger counting. Our own system of number labeling does not make clear its base-10 structure until the child can count above 20. Different linguistic expressions of number may affect initial ease of learning number concepts. Perhaps bilingual children appreciate the abstract symbolic nature of number more readily than monolingual children because there is both a commonality in the meaning of number and a difference in the way that number is expressed in different languages. The field of mathematics learning is a fertile one for examining the interplay of language and thought in development.

This volume is concerned with societal differences in methods of mathematics teaching and in expectations of mathematics learners. Our educational system has, at different times, taught arithmetic as a set of basic skills, as a conceptual system, and as a model of problem solving. Various societies teach arithmetic quite differently. Japanese parents send their children after school to abacus lessons. Using an abacus helps the child to visualize addition and subtraction and to conceptualize the base-10 system (Hatano, 1982). Similarly the Russians encourage children's reliance on finger counting and visual aids in the initial learning of addition and subtraction (Davydov, 1982). In contrast, the American school system stresses fact retrieval and discourages finger counting. The practical issue is to determine the system of instruction that is best for the child at different levels of expertise. The concern also has theoretical implications. Throughout this chapter there have been contrasting views of how the child acquires mathematical knowledge. The constructivist view characterizes the child as creator of his or her own system, which emerges with little or no instruction on the part of the parents. Mathematics is not completely self-taught but is learned through informal activities in the home and formal instruction.

Hence for developmentalists interested in the socialization of cognition, mathematics is an area in which to examine the influence of the family and the school on mathematical skill.

The influences of school and parent on a child's learning are not straightforward because the two sources of socialization may or may not complement one another. Some parents' conversations with young children are studded with test questions requiring children to retrieve facts, evaluate the accuracy of statements, and to problem-solve. Parents, particularly middle-class ones, convert daily transactions into lessons and they heighten the child's metacognitive awareness and cognitive monitoring (e.g., Hall, Scholnick, & Hughes, in press). Dinner table conversation in many homes prepares the child for the interchanges that take place in the classroom. Sigel (1981) describes a parental style of speech (distancing) that promotes abstraction by focusing the child on future imagined possibilities and relations among events rather than just the concrete details of daily life. That style of discourse may foster the skill in abstraction that is the basis for mathematics. When the parental style of interaction is different from the school's and the way of discussing concepts is different from the way the school transmits information, the child may have to learn a new speech pattern in school. Comprehending the new style places an additional burden on the child who is also mastering the conceptual content of arithmetic. When textbooks use examples drawn from the daily life of children much different from the pupils who are using the text, then the difficulty of encoding and understanding the problems is increased. Children may also differ in the frequency of exposure to informal mathematical problems and their style of handling those problems may also differ from their style of handling academic tasks. Everyday arithmetic may not completely resemble textbook mathematics. Often, adults in daily life use strategies that may minimize effort because small errors are unimportant (e.g., Lave, Murtaugh, & de la Rocha, 1984; Scribner, 1984). Parents may provide children engaged in arithmetic activities with the kind of contextual support lacking in school arithmetic. Arithmetic is an interesting and important field in which to study socialization because socialization comes from informal and formal instruction. We still lack knowledge about how parents may facilitate the bridge between daily life outside of school and academic activities through the content and style of the lessons they provide for children. Nor do we know how the parents' own "theories" of mathematics influence what and how they teach children. Parents differ in the value they place on mathematical skill and the support they provide for its acquisition.

There are also strong biases about who can learn mathematics, and pervasive individual differences in learning skills. Although girls are at least as competent as boys in learning arithmetic calculations, parents, teachers, and the girls themselves underestimate their competence (Entwisle & Baker, 1983). There is hot debate about whether there are genetic differences in mathematical capacities (Benbow & Stanley, 1980). Similarly, there may be cultural differences in the patterning of skills that reflect attitudes and values about the role of mathematics in daily life. Thus the field of mathematics can provide information about the interplay of biological and attitudinal differences in the emergence of cultural skill.

DEVELOPMENTAL PSYCHOLOGY AND THE LANGUAGE OF MATHEMATICS

The list of reasons why developmental psychologists might want to study mathematics learning is very heterogenous because there are underlying tensions in the field itself. There are many debates about the generality of cognitive structures, the development of rich conceptual systems, the evolution of abstraction, the nature of learning, and developmental continuity. Because mathematics provides a pure case for studying knowledge acquisition, mathematical theory has a logical cast, and a variety of developmental theories are similarly cast into a language that is inherently mathematical. The study of mathematics thus becomes a tool for validating those theories. These are abstract theories of an abstract cognizer learning an abstract domain. The theories are not without relevance to the educational enterprise because the analyses they provide of the content of arithmetic and the process of its mastery can influence and have influenced curriculum design.

A contrasting approach redefines learning and development by placing it within the context of socialization practices at home and in the school. It regards the content of mathematics as an ever-changing field that exists not as an abstraction but as a piece of a linguistic and cultural fabric. This tradition is heavily influenced by Vygotsky (1978). This view emphasizes individual differences in the conception of mathematics and the process of acquisition of mathematical concepts. Although mathematics is not a Rorschach blot that every society and family within a society can interpret, nevertheless there may be fundamental differences in the aspects of mathematics that different cultures may stress and in how mathematics is taught that may account in part for differences in mathematics achievement.

Children may be exposed to a variety of conflicting views about the nature and practice of arithmetic. The contextual view sets a very different research agenda for developmental psychologists and for educators. It is taken for granted that mathematics learning is embedded in a cultural context. Yet there are many cultural contexts within a society so that everyone does not approach adding and subtracting in the same way. It becomes important to describe the cognitive and motivational components of those contexts. Since children operate in many contexts (e.g., Bronfenbrenner, 1979), it is equally important to specify the necessary bridging structures between home and school and between one concept and another that enable the child to learn mathematics. There is not one mathematics, but many overlapping variations that may also give rise to many mathematics curricula.

ACKNOWLEDGMENTS

I wish to thank Susan Chipman, who during my stay at the National Institute of Education convinced me that mathematics learning was something a developmental psychologist should think about and who introduced me to the body of research that NIE was supporting.

REFERENCES

Antell, S. E., & Keating, D. P. (1983). Perception of numerical invariance in neonates. *Child Development, 54,* 695–701.

Anderson, N. H., & Cuneo, D. O. (1978). The height + width rule in children's judgments of quantity. *Journal of Experimental Psychology: General, 107,* 335–378.

Benbow, C., & Stanley, J. (1980). Sex differences in mathematics ability: Fact or artifact? *Science, 210,* 1262–1264.

Bornstein, M. H., & Sigman, M. D. (1986). Continuity in mental development from infancy. *Child Development, 57,* 251–274.

Braine, M. D. S., & Rumain, B. (1983). Logical reasoning. In P. H. Mussen (Ed.), *Handbook of child psychology: Vol. 3. Cognitive development* (4th ed., pp. 263–340). New York: Wiley.

Brainerd, C. J. (1979). *The origins of the number concept.* New York: Praeger.

Bronfenbrenner, U. (1979). *The ecology of human development: Experiments by nature and design.* Cambridge, MA: Harvard University Press.

Brown, J. S., & Van Lehn, K. (1982). Towards a generative theory of "bugs." In T. P. Carpenter, J. M. Moser, & T. A. Romberg (Eds.), *Addition and subtraction: A cognitive perspective* (pp. 117–135). Hillsdale, NJ: Lawrence Erlbaum Associates.

Case, R. (1985). *Intellectual development: Birth to adulthood.* New York: Academic Press.

Chi, M. T. H., Glaser, R., & Rees, E. (1982). Expertise in problem solving. In R. J. Sternberg (Ed.), *Advances in the psychology of human intelligence* (Vol. 1, pp. 7–75). Hillsdale, NJ: Lawrence Erlbaum Associates.

Cooper, R. G., Jr. (1984). Early number development: Discovering number space with addition and subtraction. In C. Sophian (Ed.), *Origins of cognitive skills* (pp. 157–192). Hillsdale, NJ: Lawrence Erlbaum Associates.

Davydov, V. V. (1982). The psychological characteristics of the formation of elementary mathematical operations in children. In T. P. Carpenter, J. M. Moser, & T. A. Romberg (Eds.), *Addition and subtraction: A cognitive perspective* (pp. 224–238). Hillsdale, NJ: Lawrence Erlbaum Associates.

Entwisle, D. R., & Baker, D. P. (1983). Gender and young children's expectations for performance in arithmetic. *Developmental Psychology, 19,* 200–209.

Fodor, J. A. (1983). *The modularity of mind.* Cambridge, MA: MIT Press.

Fuson, K. C., & Hall, J. W. (1983). The acquisition of early number word meanings: A conceptual analysis and review. In H. P. Ginsburg (Ed.), *The development of mathematical thinking* (pp. 49–107). New York: Academic Press.

Gardner, H. (1983). *Frames of mind: The theory of multiple intelligences.* New York: Basic Books.

Gelman, R., & Baillargeon, R. (1983). A review of some Piagetian concepts. In P. H. Mussen

(Ed.), *Handbook of child psychology: Vol. 3. Cognitive development* (4th ed., pp. 167–230). New York: Wiley.

Gelman, R., & Baillargeon, R. (1983). A review of some Piagetian concepts. In P. H. Mussen (Ed.), *Handbook of child psychology: Vol. 3. Cognitive development* (4th ed., pp. 167–230). New York: Wiley.

Gelman, R., & Gallistel, C. R. (1978). *The child's understanding of number*. Cambridge, MA: Harvard University Press.

Ginsburg, H. (1977). *Children's arithmetic: The learning process*. New York: Van Nostrand.

Ginsburg, H. (1983). Introduction. In H. P. Ginsburg (Ed.), *The development of mathematical thinking* (pp. 1–5). New York: Academic Press.

Greeno, J. G., Riley, M. S., & Gelman, R. (1984). Conceptual competence and children's counting. *Cognitive Psychology, 16*, 94–143.

Groen, G. J., & Resnick, L. (1977). Can pre-school children invent addition algorithms? *Journal of Educational Psychology, 67*, 17–21.

Hall, W. S., Scholnick, E. K., & Hughes, A. (in press). Contextual constraints on the usage of cognitive words. *Journal of Psycholinguistic Research*.

Hatano, G. (1982). Learning to add and subtract: A Japanese perspective. In T. P. Carpenter, J. M. Moser, & T. A. Romberg (Eds.), *Addition and subtraction: A cognitive perspective* (pp. 211–223). Hillsdale, NJ: Lawrence Erlbaum Associates.

Johnson-Laird, P. N. (1983). *Mental models*. Cambridge, MA: Harvard University Press.

Klahr, D. (1984a). Commentary: An embarrassment of number. In C. Sophian (Ed.), *Origins of cognitive skills* (pp. 295–305). Hillsdale, NJ: Lawrence Erlbaum Associates.

Klahr, D. (1984b). Transition processes in quantitative development. In R. J. Sternberg (Ed.), *Mechanisms of cognitive development* (pp. 101–139). New York: Freeman.

Klahr, D., & Wallace, J. G. (1976). *Cognitive development: An information processing system*. Hillsdale, NJ: Lawrence Erlbaum Associates.

Kline, M. (1980). *Mathematics: The loss of certainty*. New York: Oxford Unviersity Press.

Lave, J., Murtaugh, M., & de la Rocha, O. (1984). The dialectic of arithmetic in grocery shopping. In B. Rogoff & J. Lave (Eds.), *Everyday cognition: Its development in social context* (pp. 67–94). Cambridge, MA: Harvard University Press.

Maratsos, M. (1983). Some current issues in the study of the acquisition of grammar. In P. H. Mussen (Ed.), *Handbook of child psychology: Vol. 3. Cognitive development* (4th ed., pp. 707–786). New York: Wiley.

Piaget, J. (1965). *The child's conception of number*. New York: Norton.

Piaget, J. (1971). *Genetic epistemology*. New York: Norton.

Piaget, J. (1977). *The development of thought: Equilibration of cognitive structures*. New York: Viking.

Resnick, L. B. (1983). A developmental theory of number understanding. In H. P. Ginsburg (Ed.), *The development of mathematical thinking* (pp. 109–151). New York: Academic Press.

Riley, M. S., Greeno, J. S., & Heller, J. I. (1983). Development of children's problem-solving ability in arithmetic. In H. P. Ginsburg (Ed.), *The development of mathematical thinking* (pp. 153–196). New York: Academic Press.

Rosch, E., & Mervis, C. B. (1975). Family resemblances: Studies in the internal structure of categories. *Cognitive Psychology, 7*, 573–605.

Saxe, G. B., & Posner, J. (1983). The development of numerical cognition: Cross-cultural perspectives. In H. P. Ginsburg (Ed.), *The development of mathematical thinking* (pp. 291–317). New York: Academic Press.

Schank, R. C. (1982). *Dynamic memory: A theory of reminding in computers and people*. Cambridge, England: Cambridge University Press.

Scribner, S. (1984). Studying working intelligence. In B. Rogoff & J. Lave (Eds.), *Everyday cognition: Its development in social context* (pp. 9–40). Cambridge, MA: Harvard University Press.

Siegler, R. S., & Richards, R. S. (1983). Development of two concepts. In C. J. Brainerd (Ed.), *Recent advances in cognitive developmental theory: Progress in cognitive developmental research* (pp. 51–121). New York: Springer-Verlag.

Siegler, R. S., & Robinson, M. (1982). The development of numerical understandings. In H. W. Reese & L. P. Lipsitt (Eds.), *Advances in child development and behavior* (Vol. 16, pp. 242–312). New York: Academic Press.

Siegler, R. S., & Shrager, J. (1984). Strategy choices in addition and subtraction: How do children know what to do? In C. Sophian (Ed.), *Origins of cognitive skills* (pp. 229–293). Hillsdale, NJ: Lawrence Erlbaum Associates.

Sigel, I. (1981). Social experience in the development of representational thought: Distancing theory. In I. E. Sigel, D. M. Brodzinsky, & R. M. Golinkoff (Eds.), *New directions in Piagetian theory and practice* (pp. 203–217). Hillsdale, NJ: Lawrence Erlbaum Associates.

Starkey, P., Spelke, E., & Gelman, R. (1982). *Detection of intermodal numerical correspondence by human infants.* Paper presented at the meeting of the International Conference on Infant Studies, Austin, TX.

Strauss, M. S., & Curtis, L. E. (1984). Development of numerical concepts in infancy. In C. Sophian (Ed.), *Origins of cognitive skills* (pp. 131–155). Hillsdale, NJ: Lawrence Erlbaum Associates.

Van Lehn, K. (1983). On the representation of procedures in repair theory. In H. P. Ginsburg (Ed.), *The development of mathematical thinking* (pp. 197–252). New York: Academic Press.

von Glasersfeld, E., & Richards, J. (1983). The creation of units as a prerequisite for number: A philosophical review. In L. P. Steffe, E. von Glasersfeld, J. Richards, & P. Cobb (Eds.), *Children's counting types: Philosophy, theory and application* (pp. 1–20). New York: Praeger.

von Glasersfeld, E., Steffe, L. P., & Richards, J. (1983). An analysis of counting and what is counted. In L. P. Steffe, E. von Glasersfeld, J. Richards, & P. Cobb (Eds.), *Children's counting types: Philosophy, theory and application* (pp. 21–44). New York: Praeger.

Vygotsky, L. S. (1978). *Mind in society.* Cambridge, MA: Harvard University Press.

Wilkinson, A. C. (1984). Children's partial knowledge of the cognitive skills of counting. *Cognitive Psychology, 16,* 28–64.

Chapter 6

Patterns of Experience
and the Language of Mathematics

Manon P. Charbonneau and Vera John-Steiner
University of New Mexico

The examination of the quantitative proficiency of minority students provides a window on the difficulties many young people have with the way mathematics is currently taught in the schools. In truth, it provides a more drastic picture of those aspects of mathematics which Underhill (1985) clearly describes as evidence that "we are teaching many mathematical concepts and skills that many learners cannot grasp and some major ones that most of the learners in many classrooms cannot understand" (p. 2). The Carnegie Forum's report (1986) on the condition of American education strongly states that "it would be fatal to assume that America can succeed if only a portion of our school children succeed."

Demographically, this report also points out that by the year 2000, one out of every three Americans will be a member of a minority group. The success of these children with the educational process, and especially in the fields of mathematics and science, is clearly a critical and crucial problem. We address two issues that affect the minority student in significant ways: The role of culturally patterned activities in children's cognitive development, and the languages of mathematics.

THE PATTERNS OF EARLY EXPERIENCE AND CULTURAL INFLUENCES ON NUMERICAL DEVELOPMENT

Influences of Family Life and Environment on Learning Style

In every culture, children engage in activities that have an underlying logic for numerical and visual-spatial relations. Many of these numerical referents are accessed differently because of cultural patterns. For instance, members of rural communities have a different and often more contextually relevant sense of space and distance than those raised in urban majority culture (Bland, 1974; Connelly, 1979; Wolcott, 1982). Seasons and cycles guide the structure of activities, rather than the clock-time that dominates most of our lives. Some of the patterns of family life in minority cultures are also quite different from that of the majority. The socialization process for Hispanic children, as well as native Americans, is one of cooperation and sharing rather than competition with the focus on individual achievement commonly seen in the majority culture. Delgado-Gaitan and Trueba (1985) have pointed out that the cooperative attitudes developed in the homes of the Hispanic communities they studied directly affected the children's understanding of "shared work." Although their research also found that the specific Mexican-American children in their study were highly versatile in integrating the values of the home with those of the school, this is not always the case. Many native American children, for example, have great difficulty with the direct questioning techniques employed in most school situations. Philips' research on the Warm Spring Reservation (1972) emphasized that native American children preferred to interact with other children rather than with adults. This preference for quiet peer interaction is underscored by many native American parents who also are uneasy in parent/teacher or parent/school meetings, especially if the topic under consideration is in any way controversial or confrontational. Direct questioning techniques rarely yield answers that truly reflect the thinking or feelings of these parents.

Learning by doing (direct experience) and careful observation of peers and elders at work is a common practice among native American children. In a similar vein, native American children have strong visual-spatial skills and they gain significantly from learning experiences that utilize their observational strength (John-Steiner & Ostereich, 1975). These children's relatively high performance on drawings (in contrast to non-native American children) provides further support for the role of graphic skills as a significant channel of communication and representation. Mathematics programs for young children that encourage the use of drawing the results of numerical experiences (such as weighing and measuring) have been a highly successful means of teaching mathematical representation.

The research of Au (1980) shows the importance of understanding the participant structure of verbal interactions in ethnic minority children's lives. Hawaiian children in the Kamehameha Early Education Program (KEEP) were quite successful in early reading tasks through the use of informal group interaction; this group interaction allowed and encouraged the children to interrupt the lesson when they had something to contribute to what was being discussed, either by the teacher or another child. Au points out that this informal interaction has been educationally successful for children with Hawaiian backgrounds because it resembles the culturally patterned story-telling practices of Hawaiian adults.

Valverde (1984) cites the work of Casteneda and Ramirez on cognitive styles of Hispanic children, which proposes a specific learning style that is more field sensitive than that of Anglo children. The sources of such holistic organization are linked, by these researchers, to large families and children's need to work together in their community setting. DeAvila (this volume) describes highly successful work in mathematics on the part of Hispanic children who were encouraged to work cooperatively in small groups, interacting verbally with each other and with effective physical materials, as well as with the teacher. Working together rather than individually significantly enhances the achievement of these children. Cultural teaching/learning practices vary not only from ethnic group to ethnic group, but also within groups, so care must be taken not to assume that practices that work well with one group of children will necessarily benefit and be successful with another. Careful testing of culturally sensitive teaching methods are needed before they can be widely adapted.

Games, Play, and Shared Activities

Many of the basic understandings and use of number that children have developed before entering school stem from games and recreational activities indigenous to the culture. It is doubtful that there is any ethnic group that does not have a multitude of childhood games and activities that, if not number oriented, at least depend upon numbers for scoring. KUI TATK, a newsletter of the Native American Science Education Association, carried an insert depicting the "Mathematics of Native American Games of Chance" complete with a map showing which native American groups enjoy these games and regional variations of them (Winter issue, 1986). This publication features articles on learning styles of native American children and is of immediate value to teachers of these children who are not from the majority culture.

It is also true that the young child who explores the roundness of a ball or the varied shapes of building blocks is experiencing one of his/her first geometric discoveries, underscoring the direct, hands-on manipulation approach to learning that characterizes young children of all cultures. Besides the more organized use of numbers in games, children everywhere, in play and in their contributions to family activities, combine, take part, share, and make comparisons. Although oversimplified, these activities certainly constitute an early view of mathematics and provide for a practical sense of the use of numbers in the everyday, "real" world. Children's participation in the family routines of work constitutes, in many cultures, a basis for survival and subtly underlines many important fundamental concepts of mathematics. The sorting of harvest crops in some native American communities certainly contributes to the children's ability to categorize and classify in the early stages of representational development. It has been stated that Hopis grow 24 varieties of corn and 23 varieties of beans, and of the 150 species of wild plants available on the mesas and desert floors, the Hopi use 134 of them in their daily life (Kennard, 1979). The direct experience Hopi children have with this sorting of the harvest and estimating quantities provides a strong background for some very basic mathematical concepts.

Farming communities and families in general provide rich experiences with spatial concepts and estimating; of distance, such as depths for planting seed and harvest yield: "make the hole about so deep;" "put about 10 seeds in each hole;" "this sack of seed will plant the entire field," are examples of relevant adult directives. Children who have direct, extensive experience with activities such as these seem to acquire an overall "sense" of number that enhances their basic understanding. And considering the emphasis that the National Council of Teachers of Mathematics (1980) is putting on the area of estimation and approximation, children who grow up with a strong background in the use of these two concepts can benefit from their informal experiences. One example of bringing formal and informal experiences is provided by a Hopi teacher (Honyouti, personal communication, 1984). He understood the strengths of his students in estimation skills and built upon these to such an extent that these native American learners did outstanding work in mathematics even after they had completed their schooling on the reservation. These learners experienced a gradual shift from informal mathematical thinking to the more formal approaches demanded by text books described by Ginsburg and Russell (1981).

In many cultural and ethnic groups, children's contribution to the daily subsistence chores provides equally rich and fundamental experiences that establish basic understandings of mathematical relationships, and provide the scaffolding necessary for more formal study. The work of Ginsburg and Russell (1981) supports the theory that children from all ethnic and language minorities enter school with good backgrounds, but schooling alters this.

Making Connections Between Formal and Informal Learning
Experiences

Children everywhere experience the difficulty of making connections between learning informally, in the context of the home situation, and in the decontextualized learning environment of school. However, such a difficulty of bridging their own childhood understanding and use of number to the study of formal school arithmetic seems to be particularly serious for minority children (Cole, Gay, Glick, & Sharp, 1971; Gay & Cole, 1967; Saxe, this volume). Scott (1986) cites work in the field of "ethnomathematics" by Hunting (1985), in which Hunting proposed a research program aimed at identifying activities and processes employed in other cultures that "have potential for connections with mathematical concepts, techniques and procedures."

Thus, the use of these practical strategies when solving problems seems to help the child make sense of the very specific world and vocabulary of mathematics encountered in the classroom. Leap (this volume) cites the "practical approach" as one of the strategies used by the Ute children in his study. Ginsburg (1977) states that failure to help children make these "practical" connections is one of the serious defects in the teaching of elementary arithmetic and mathematics in the schools. Others, by contrast, point out that "practical" language often does not map onto the "technical" language of mathematics, and, therefore such learning strategies of applying home experience to school problems is less than satisfactory. Clearly such learning styles need to be meshed with concrete experiences that help map the mathematics/language relationships (Spanos et al., this volume).

How children understand numbers, whether in the context of informal learning or the more formal mathematics of the classroom, hinges very critically on language—their own and the highly specific language of the subject itself. Therefore, because language is the critical mediator of concept formation and concept development, we devote the second part of our chapter to this important concept.

THE LANGUAGE OF MATHEMATICS

Mathematics speaks to children in many different ways. It "speaks" through its very specific vocabulary, of course, but also through the broad and varied types of numerical problem solving placed before the student; from very simple counting to the more formal concepts and relationships of arithmetic, algebra, geometry, and other specialized branches of mathematics. Children acquire a variety of spontaneous concepts through home activity and playful encounters with their peers. However, it is difficult to make the connection with decontextualized learning without sensitive assistance, a point make by Cocking and Chipman (this volume). In our discussion of the languages of mathematics, we limit ourselves to these areas of mathematics encountered by preschool and elementary age children, namely, basic arithmetic.

Counting and the Count-Word Sequence

Young children's first experiences with numbers seem to stem from learning to enumerate; both the count-word sequences and the understanding of one-to-one matching relationships between the word and the objects being counted. Children everywhere struggle to acquire these two very basic concepts, and those children who enter school with both the count-word sequence and the cognitive understanding of one-to-one correspondence and the concept of number clearly have an advantage when confronted with the direct instructional sequence of formal school.

However, for those children whose basic experience with number may have been with a counting system that is unlike that of majority students (and school), just making the shift to the formal system can be a monumental conceptual task. Saxe (this volume) carefully explains the Oksapmin body-part counting method, and the efforts made by these children in school to adapt that system to school arithmetic with varying degrees of success. He also cites the linguistic regularity of Japanese numeration that cannot help but support the young Japanese child's understanding of the structure of the base-10 system of numeration. Children may learn other count-word sequences in their native language that might have been created, historically, around a base system other than 10. Some native American languages, as well as others, linguistically indicate this occurrence, and there is considerable literature to indicate that native American, as well as black children, have an inordinately difficult time with understanding symbolic numeration, perhaps for this language-related reason. It is unlikely that these children immediately profit from direct instruction and lessons of the majority classroom without some carefully thought-out experiences that help to span this gap of language discontinuities. English names for teen numbers certainly do nothing to clarify the child's understanding of "10" as "one group of ten and no extra units, 11 as one group of ten and one unit," and so forth. Further, the verbalization of these teen numbers adds a further dimension of difficulty to the problem of writing these numbers correctly. Quite logically, the young child could (and often does) deduce that "thirteen" should be written with the "3" first! These types of errors are not restricted to children of language minorities but are seen frequently in primary classrooms nationwide. Poems and rhymes may help children become familiar with different ways of expressing numbers: "4 and 20 blackbirds," "four score. . . ." Although we do not either speak or count in this manner, familiarity with such expressions (and their meanings) could conceivably help some children. Another part of the challenge of conceptualizing numerical relationships is that of transforming the child's experience with spatial relationships into the sequential languages of counting and of syntax.

Arithmetic and the Concept of Number

Children facing direct instruction as school beginners must understand the count-word sequence and one-to-one correspondence as well as the meaning of "numberness;" what is "three," "four," "five" and so on? Piaget has left us with an incredibly rich background of the child's concept of number and especially the developmental growth of two critical understandings: ordinal and cardinal number. The assumption that if a young child can count, he/she must certainly understand "twoness," "threeness," and so forth, is shattered daily in classrooms everywhere.

One of the authors recently watched a 4-year-old, obviously very bright and capable, when asked to count 10 pairs of scissors to be returned to another classroom; this student carefully counted "1" (placing a pair of scissors to the side), "2" (placing another pair with the first), "3" (again putting a third pair with the others) and when reaching "10," with 10 pairs of scissors carefully placed together, went right on: "11" (placing another pair of scissors), "12," and so on. His teacher quietly placed a hand over his, and said, "No, dear, TEN pairs of scissors. Let's start again." The child repeated his performance, this time reaching "14" before being stopped and told to start again. The third time, when he reached 10 and again continued, the teacher suggested he stop and go play, turned to her colleagues and wondered aloud, "How can he count with a one-to-one correspondence and not know when he reaches 10?" It did not occur to any of the adults present to use their two hands, with fingers extended on the table, and ask the child to lay out as many scissors as there were fingers, in a direct one-to-one physical matching as an experience with quantity. His previous experience was obviously restricted to rote counting with a need for each word to match an object being counted.

Herscovics (in Steffe, Von Glasersfeld, Richards, & Cobb, 1983) comments that teachers working with young children need "to be jolted back into awareness of the difficulties in learning the number concept" (p. 13). He cites a teacher-training experiment in which the first 10 numbers in English were replaced by the words "pauve, petit, patineur, prends, patience, pour, pouvoir, patiner, plus, paisiblement." He describes the teacher reaction by indicating that the semantic content of the sentence made for quick memorization; however, the use of the words for simple arithmetic tasks produced great difficulties. Counting backwards was exceedingly hard work, and trivial addition facts became major problems requiring counting on the fingers while repeating the words several times. Herscovics states "that experiences such as these indicate the need in didactics to go beyond a purely mathematical analysis of a concept and to consider it from an epistemological point of view" (p. 13). The conclusion is that children's logic follows a different line from mature adult taxonomy. Only by weaving together how the child conceives number (namely, the way the child structures base-10 numbers) with the scientific concepts of adult can the teacher fully assist the young learner. Teachers, indeed, need to know more about children's developing epistemologies.

We stated previously that Leap (this volume) cites an often used strategy by Ute children of reverting to practical processes when confronted by a problem involving an operation with which they had only limited experience (i.e., multiplication and division). Somehow, encouragement of such a strategy, which seems so logical for all children, is limited. Even in simple arithmetic, a strict interpretation of the vocabulary of the subject, as well as use of the accompanying correct symbolic representation, is what is asked of and expected of children right from the beginning. As Ginsburg (1977) suggests, we need to make a point of helping children relate the world of school arithmetic to the real world of numbers that makes up the background with which they enter the classroom. It is critical that children have the opportunity to develop the cognitive uses of language, the language of school and specifically mathematics, which combines the codification of *spontaneous concepts* and *scientific concepts*. The latter is acquired in formal learning situations and is dependent on the more academic form of language (Vygotsky, 1962). The specific vocabulary and symbolic representation required in the study of school arithmetic must be acquired, but perhaps a more carefully guided process of acquisition, which includes, initially, the use of "practical processes" and language as well as child-invented algorithms might help bridge this immense gap.

Many children, and especially minority children, seem to go through the elementary years feeling that numbers "control" them, instead of their having power over numbers! And with such a background, it is understandable that they shy away from the advanced mathematics of high school and beyond. In the case of school arithmetic, it would seem that there is definitely a subordination of learning to teaching; a priority of subject matter over how children learn.

THE SHIFT FROM TEACHING TO LEARNING

Instead of using the valuable knowledge we already possess about how children learn spontaneously, with appropriate guidance and verbal interaction, emphasis is still being put on skills proficiency. Such proficiency is taught in an inflexible sequence accentuating drill and practice rather than developing the underlying cognitive processes that guide the acquisition of number concepts.

Children simply do not learn the same things in the same way, even within the same cultural background. Specifically in the domain of mathematics, children construct their understanding of number over many years, slowly and carefully, fitting new experiences into current cognitive structures. Many children, and especially those from ethnic minorities, often misinterpret what is being asked of them by teachers (see Mestre, this volume). Steffe et al. (1983) have emphasized the need for teachers and researchers to watch and see what it is that children do with the tasks asked of them in order to determine the intentions and understandings of the child, clearly the subordination of teaching to learning.

If the shift from teaching to learning is to occur (and the need for this shift is underscored in the report of the Carnegie Forum), the role of the teacher will drastically change. Instead of being a *director,* the teacher of young children will become an *orchestrator* of appropriate experiences, with physical materials and language. The teacher will be more of a catalyst to learning rather than a *presenter* of a body of facts and figures. Teachers who have already assumed such a role have become very effective "kid watchers," using these careful observations of children interacting with materials, their peers, and their teachers, as cues and clues to concept formation and the need for planning the "next step."

The learning of mathematics is a complex process for all children. Successful study of this important discipline must take into consideration the many and varied background experiences with which children come to school. Strengths of learning through the visual, observational mode, and the tactile, manipulative mode need to be utilized. Some children may need direct instruction, whereas others may not have engaged in this type of verbal interaction. Some children learn better in small groups or pairs, but others thrive in more self-determined learning situations, and there are also some children who do profit from the traditional large group instructional setting.

If *all* children, regardless of ethnic origin or language background, are to be successful in the skills of mathematics, we must be prepared to support differing learning styles and a more flexible curriculum content. For the minority student, this means that classroom experiences need to be beyond textbook-based instruction. Hands-on, experiential learning, critical to the learning of all young children, needs to be structured to reflect cultural practices in the child's home and neighborhood, emphasizing interactional routines that characterize different community practices.

The challenge of mathematical thinking and its availability to children is beautifully rendered by the educational philosopher, David Hawkins (in prep.). He tells the following story:

> There is a charming story about a waif from the streets of London whose young friend and mentor finds in her a precocious interest in numbers. After a considerable time of involvement with the subject she reveals a deep concern about the nature of numbers, constructing a remarkable model to explain them. Casting the shadow of some familiar object on a screen, she then insists on a cardboard replica of the shadow. Turned on edge, its shadow is a thin straight line segment. This in its turn is represented by a thin stick, which when held normal to the screen casts a small dot-shadow. This, she announces, is how Mr. God makes numbers. All the complexity of real objects is reduced by this three-fold shadowing to their least common character, each distinguishable from the others but counting nevertheless, counting as one among a number. (p. 50) The rules of matching and counting, abstracted from our dealings with nature, yield more than the specific items or experience which went into the abstracting of them. The system of number standards and of procedures for generating new ones has, therefore, the nature of a world which transcends our knowledge of it and even our conjectures about it: complex, surprising, and inexhaustible. (p. 99)

REFERENCES

Au, K. (1980). Participation structures in a reading lesson with Hawaiian children: Analysis of a culturally appropriate instructional event. *Anthropology and Education Quarterly, 11*(2), 91–115.

Bland, L. (1974). *Visual perception and recall of school-age Navajo, Hopi, Jicarilla Apache and Cuacasian children of the Southwest.* Unpublished Ph.D. dissertation, University of New Mexico, Albuquerque, NM.

Carnegie Forum Task Force on Teaching as a Profession (1986). *A nation prepared: Teachers for the 21st century.* New York: Carnegie Forum on Education and the Economy.

Cole, M., Gay, J., Glick, J., & Sharp, D. W. (1971). *The cultural context of learning and thinking.* New York: Basic.

Connelly, J. C. (1979). Hopi world view. In A. Ortiz (Ed.), *Handbook of North American Indians (Southwest).* Washington, DC: Smithsonian Institute.

Delgado-Gaitan, C., & Trueba, H. (1985). Dilemma in the socialization of Mexican-American children: Sharing for cooperation and copying for competition. *Quarterly Newsletter of the Laboratory of Comparative Human Cognition, 7,* No. 3.

Gay, J., & Cole, M. (1967). *The new mathematics and an old culture: A study of learning among the Kpelle of Liberia.* New York: Holt, Rinehart & Winston.

Ginsburg, H. (1977). *Children's arithmetic: How they learn it and how you teach it.* Austin, TX: Pro:Ed.

Ginsburg, H., & Russell, R. L. (1981). Social class and racial influences on early mathematics thinking. *Monographs of the Society for Research in Child Development, 46* (6, Serial No. 193).

John-Steiner, V., & Ostereich, H. (1975). *Learning styles of Pueblo children.* Internal Report, College of Education, University of New Mexico, Albuquerque, NM.

Kennard, E. (1979). Hopi economy and subsistence. In A. Ortiz (Ed.), *Handbook of North American Indians (Southwest).* Washington, DC: Smithsonian Institute.

National Council of Teachers of Mathematics. (1980). *An agenda for action: Recommendations for school mathematics of the 1980's.* Reston, VA: National Council of Teachers of Mathematics.

Native American Science Education Association. (1986, Winter). Mathematics of Native American games of chance. *KUI TATK.*

Philips, S. V. (1982). Participant structures and communicative competence: Warm Springs Indian children in community and classrooms. In C. Cazden, V. John, & F. Hymes (Eds.), *Functions of language in the classroom.* New York: Teachers College Press.

Scott, P. (1986, September). Review of *Learning, aboriginal world view and ethnomathematics* by Robert P. Hartung. "Have You Seen. . . ." *International Study Group on Ethnomathematics,* (2), *1.*

Steffe, L. P., Von Glasersfeld, E., Richards, J., & Cobb, P. (1983). *Children's counting types: Philosophy, theory and application.* New York: Praeger.

Underhill, R. (1985). Let's diagnose the child AND the curriculum. *Arithmetic Teacher, 33*(4), 1.

Valverde, L. A. (1984). Underachievement and underrepresentation of Hispanics in mathematics and mathematics-related careers. *Journal for Research in Mathematics Education, 15,* 123–133.

Vygotsky, L. (1962). Thought and language. Cambridge, MA: MIT Press.

Wolcott, H. (1982). The anthropology of learning. *Anthropology & Education Quarterly, 13*(2), 88.

Chapter 7

Bilingualism, Cognitive Function, and Language Minority Group Membership

Edward A. De Avila

DeAvila, Duncan and Associates, Inc.

San Rafael, CA

INTRODUCTION

The study of childhood bilingualism may be described as having both theoretical and applied importance. With respect to theoretical interests, researchers have studied the relationship between bilingualism and various cognitive processes such as intellectual development and cognitive style. On the applied side, researchers have used results from theoretical studies to design and test hypotheses regarding the effectiveness of different treatment approaches such as those found in bilingual education. The purpose of this chapter is to discuss a number of issues related to both theoretical and applied interests in the intellectual and social functioning of bilingual children in school. The chapter is presented in two major sections. In the first, we make a few introductory comments regarding the current study of bilingual students. In the second section, we review the results of a number of studies that have addressed a number of important theoretical and applied questions within the context of an educational program designed to accommodate the linguistic and educational heterogeneity of ethnolinguistic minority students.

BACKGROUND OF STUDIES

A substantial number of sources have documented the poor academic performance of language minority students in the United States. Similarly, a good number of researchers have attempted to explain this poor performance. Of particular interest to the present discussion is the widespread belief that somehow language minority students are at an academic disadvantage by virtue of intellectual, verbal, motivational, and cognitive style differences that have been equated with bilingualism. Unfortunately, studies offered in support of this contention have tended to confuse both poverty and ethnolinguistic group membership with linguistic proficiency. Duncan (1979) reviewed over 100 studies on the effects of bilingualism conducted in the United States over the past 50 years and found that in only four cases was the actual extent of "bilingualism" assessed. With rare exception, subjects were grouped on the basis of ethnicity, assuming linguistic proficiency.

A concrete example of this confusion is found in a recent article where Mestre, Gerace, and Lochhead (1982) found that "even though the bilinguals (i.e., "balanced") were . . . nearly equivalent in both Spanish and English, the level of proficiency in either language for the bilingual group was substantially below the level of the monolingual group in English" (p. 6–9). In other words, if bilingualism means proficiency in both of the two languages at least at the same level as that of the typical monolingual student, then this group of "Hispanic bilinguals" was not bilingual. The review of problems associated with the definition and measurement of bilingualism by De Avila and Duncan (1980) cites numerous examples of this type in a variety of research contexts. In practical terms, it is impossible to discuss the effects of bilingualism within the United States without considering language minority group membership.

Another example of confounding group membership with linguistic characteristics is revealed in the following logic. Mexican-American students are found to be more "field dependent" than Anglo counterparts (Buriel, 1975; Ramirez, 1973; Sanders, Scholz, & Kagan, 1976). To the extent that Mexican-American students are assumed to be bilingual, obtaining cognitive style differences are equated with linguistic differences. Thus, bilinguals come to be viewed as field dependent. In a review of the literature on cognitive style, De Avila and Duncan (1980) concluded "virtually no studies involving Spanish language background children have controlled for language proficiency in either Spanish or English or for intellectual development. Given the failure to control for these potentially important variables, any differences with respect to groups must remain equivocal" (p. 126).

Using an approach based on the earlier work of De Avila, Havassy, and Pascual-Leone (1976), Duncan (1979), Duncan and De Avila (1979), and De Avila, Duncan and Ulibarri (1982) found that when differences in linguistic proficiency were controlled through assessment in both languages, differences in cognitive style were to a large extent test-specific with only fully proficient bilingual students demonstrating consistent differences (higher levels of cognitive development on both Witkin and Piaget type tasks). These results replicate a number of similar studies conducted over the past 15 years which have shown that when linguistic and test-demand characteristics are controlled, many of the reported ethnolinguistic group differences in cognitive functioning fail to emerge (see De Avila & Duncan, 1980, for a review).

Finally, most ethnolinguistic group comparison studies have been motivated by an attempt to understand the impact of group variation on school performance. Findings between group differences on various cognitive tasks such as the Children's Embedded Figures Task (Goodenough & Karp, 1961) are understood to explain differences in academic achievement (Chan, 1983). In an extensive 3-year cross-cultural study involving over 1,200 students from six different ethnolinguistic backgrounds, De Avila, Duncan, Ulibarri, and Fleming (1979) found that cognitive style differences contributed less than 10% to the total variance in predicting school performance of language minority students. On the other hand, linguistic variation accounted for as much as 70% of the total variance. Cognitive style and intellectual development variables accounted for significant proportions of the variance in predicting school performance only for mainstream Anglo students.

As a result of our own studies and review of various approaches to the study of the relationship of school achievement and such factors as bilingualism or linguistic proficiency, intellectual development, and cognitive style, we have concluded that researchers would gain by considering the issue of treatment and environmental circumstances before attributing differences in school performance to either "bilingualism" or "cognitive style" variation. In this connection, we would do well to consider the underlying theory that guides our designing of programs. Finally, it can be argued that understanding the effects of bilingualism on cognitive style (or vice versa) cannot be achieved without studying the contexts and processes by which it develops. Here, De Avila and Duncan (1979) argue that cognitive style and bilingualism develop in a fashion that is indistinguishable from that of intelligence.

The major policy focus for bilingual education in the United States since the passing of the 1968 Bilingual Education Act (Title VII, ESEA) has been one of compensatory assistance. Most educators have held the view that the difficulties faced by language minority students result from lower intellectual levels of development assumed to be associated with bilingualism, cognitive style differences, motivational deficiencies, and a host of factors usually referred to under the rubric of socioeconomic status (SES) (see Rosenthal, Milne, Ginsburg, & Baker, 1981). By means of increased resources associated with the remediation of English language deficiencies through English as a Second Language programs (ESL), it is expected that children will acquire sufficient English language skills to participate fully in English-only classrooms. This perspective has been coupled with an increased recognition of the demands for equity in educational opportunity (Lau v. Nichols). Similarly, increasing and shifting demographic patterns within the U.S. population have encouraged a consideration of the increasing numbers of children who come from homes where English is not the primary langauge. As a result, American educators have become increasingly aware of the special demands placed on the schools by ethnically and linguistically heterogeneous students.

Unfortunately, however, recent reviews of the content of bilingual programs suggest an even greater emphasis on programs that foster a dependence on predetermined approaches to problem solving than is found in regular classrooms (Arias, 1982; Laosa, 1977; Nieves-Squires, 1983). Clark and Peterson (1976) report that teachers are concerned mostly with the factual information in their lessons. Other research (Bellack, Kliebard, Hyman, & Smith, 1966) suggests that teachers use extremely simple logic in their interactions with students. This simplified logic is particularly evident in bilingual classrooms in the "modified speech" patterns used by English-speaking teachers when interacting with students of "limited English proficiency" (Takahashi, 1982). Not surprisingly, one finds that current classroom practices focus on the rote learning of facts to the exclusion of more complex forms of information processing. Taba (1966) argues that the main reason for the low status of "thinking" in American classrooms, in general, and in bilingual classrooms in particular, derives from the belief that thinking is predicated on learning a body of factual knowledge, i.e., one has to know content before one can think. With respect to language minority students, English language proficiency becomes an additional prerequisite to thinking. The observation that teachers face a difficult task is underscored by the experience of any practitioner who has worked in classrooms, continually confronted with complex classroom management problems that tend to subordinate thinking to content. Thinking is difficult to foster in any classroom, let alone in a linguistically heterogeneous bilingual (or even multilingual) environment.

Organizational sociologists (Cohen, Deal, Meyer, & Scott, 1976; Intili, 1977; Perrow, 1967) point out that difficult tasks require complex support systems. To organizational sociologists and psychologists alike, the "teaching" of thinking is a much more complex operation than is normally found in the traditional classroom. This is particularly the case where teachers have little understanding regarding the nature of the learning process and have been encouraged to treat all content as the same (Berliner & Rosenshine, 1978; De Avila & Cohen, 1983), as if it were acquired by the same process. Moreover, many approaches to the teaching of content encourage routine at the expense of the more complex processes required for concept formation. Shavelson and Stern (1980) have analyzed a good deal of teacher behavior as a strong adherence to routine. According to De Avila and Cohen (1983), more individually relevant materials, more staff, and most important, more interaction among participants are required to adequately implement the processes underlying concept formation (see Bourne, 1966). Thus, fostering the development of "thinking skills" would require, according to these writers, more complex interdependent staff arrangements. Most teachers (let alone aides) have not been trained for these more complex arrangements and

most administrators whose training was based on the theory of "cultural depriva-
tion" (see Riessman, 1962) are not accustomed to them.

From a psychological point of view, it appears that the curriculum in most
classrooms requires recall skills for factual information and little else. The disadvantage of this approach, for all children, is that it limits the commitment to
individual growth and self-development espoused in the goals of "good teaching" (Intili & Flood, 1976) and actively discriminates against those language
minority students who do not share in the content embodied in the curriculum
(De Avila & Havassy, 1975). In this way, language minority students, to the
extent they are seen as "bilingual," are denied access to programs that go
beyond paired associate or rote learning (see De Avila & Cohen, 1983).

In the following, we describe a number of studies covering a number of
different topics related to the treatment of language minority students. Since the
studies are all based on the same population, we begin with a description of the
students. This description is followed by a description of the instrumentation to
collect the data.

SAMPLE

The classrooms selected for this research were part of a larger group receiving
bilingual instruction under a grant from the U.S. Office of Education (Title VII).
The Title VII program provided remedial services in both English and Spanish
with an emphasis on the basic skills. Children participating in the Finding
Out/Discubrimiento program (FO/D) were approximately 253 second-, third-,
and fourth-grade students drawn from a population of students living in suburban
and metropolitan areas surrounding San Jose, California. A total of nine classrooms from nine different schools were involved. More detailed descriptions of
the sample can be found elsewhere (De Avila, 1981). For the present it is
sufficient to note that the students making up the sample were predominantly
from lower middle-class Mexican-American backgrounds. Although there were
several black and Asian-American students the group as a whole was widely
heterogeneous with respect to linguistic proficiency in both English and Spanish.

In addition to the students who participated in the FO/D program, 300 other
students who were part of the bilingual program were used to constitute a comparison group. It should be noted that the term *comparison group* has been used
as opposed to *control group* because this was a field study and not an experiment
in the traditional sense. For example, students were not randomly assigned to
treatment and control groups (see Wilson, De Avila, & Intili, 1982, for a more
detailed discussion on this issue as it relates to the present study). The comparison group was quite similar to the study in all other respects save exposure to
the science and math activities described below.

DESCRIPTION OF TREATMENT APPROACH (FINDING OUT/DESCUBRIMIENTO)

The specific classroom or treatment activities were taken from a bilingual science and math program entitled Finding Out/Describrumento (De Avila & Duncan, 1979, 1982). FO/D is designed for use with small groups and is made up of approximately 150 activities that require measuring, counting, estimating, grouping, hypothesizing, analyzing, and reporting results. Basic skills are presented in the context of activities. For example, there are numerous situations in which students are asked to make estimations and then to check the accuracy of their estimations through subtracting estimated values from obtained values. The activities require cognitive operations found in problem-solving tasks used in research in cognitive psychology (see Lochhead, 1985; Sternberg, 1981; Valett, 1978) Individual activities are organized and presented in much the same way as in a laboratory concept-formation or learning set experiment. Thus there are a great number of trials or activities requiring the use of the same concept while the irrelevant aspects or dimensions are varied. In this way, for example, students are exposed to various aspects of the concept of number in at least 12 different activities, each presented in slightly different configurations while the "principle of solution" is held constant (see Hunt, 1961).

Students were required to work in small groups to increase verbal interaction. By presenting all material in both languages, an attempt was made to make the concepts underlying each task invariant with respect to language. Students were left free to interact in whatever language or combination of languages needed in order to facilitate communication. Linguistic differences were viewed as but one more way in which concepts could be repeated without changing the underlying concept (i.e., principle of invariance). All students were asked to complete each activity and a corresponding worksheet (also provided in both languages). Students worked in small groups where they were free to interact or discuss each activity as well as to assist one another in filling out the worksheets. Student flow was managed by the teacher who also facilitated understanding of the task instructions, which were provided in cartoon format with text in both English and Spanish. The child could thereby use the multiple resources provided by the pictograph, text, teachers, or peers.

OUTCOME MEASURES

Student progress and behavior were assessed at different points during the 14 weeks required to cover all of the activities. A variety of commercially and specially developed instruments were used. Taken as a whole, the data collected were based on both direct and indirect assessment methods including (a) paper-

and-pencil tests, (b) classroom observations, (c) daily performance on worksheet assignments. Results are reported in two sections.

Paper-and-Pencil Outcome Measures

Three paper-and-pencil measures were administered on a pre/post basis to assess learning outcomes. Their content and administration are described here.

Intellectual Development. The Cartoon Conservation Scales (De Avila, 1980; De Avila & Pulos, 1978; Fleming & De Avila, 1980) are a collection of Piagetian-inspired tasks that are group administered by means of a cartoon booklet. In all there are 32 items that fall into the eight subscales including (1) number, (2) length, (3) substance, (4) distance, (5) horizontality, (6) egocentricity, (7) volume, and (8) probability. The concepts embodied in the CCS more closely resemble various definitions of thinking skills than any other of the measures in the test battery. However, insofar as the concepts underlying the CCS are not directly covered as part of the activities, they should be viewed as measures of the extent to which concepts covered in the activities "generalized" to broader mental processes. See Bruner, Oliver, and Greenfield (1966) for a more general discussion between conservation tasks and "thinking skills."

Academic Achievement. The Comprehensive Test of Basic Skills (CTBS) is a nationally norm-referenced test of school achievement. It is made up of two sections, math and reading. For the present purposes, it is important to bear in mind that the math section is broken down into three subscales: computation, concepts, and applications. The items that make up these subscales differ from one another not only in content but in the methods used to teach them. See De Avila and Cohen (1983) for a more detailed analysis of these differences. For example, math computation is routinely taught through memorization, rote, or paired associate learning. On the other hand, math concepts items more closely resemble concepts found in the literature on concept formation (see Bourne, 1966). It was expected that the FO/D experience would have a stronger impact on Math Applications than on subscales requiring less in the way of "thinking skills," i.e., math computation.

Science-Math. The MINI test was designed (see Hansen, 1980, for a description of its development and psychometric properties) as a content-referenced multiple-choice test of the vocabulary and concepts covered in the activities. It was presented in both pictorial and written formats in an attempt to reduce the effects of reading skills. The test was administered in either English or Spanish versions depending on the proficiency of the student. Scores are reported as the total number of items correctly answered divided by the number attempted. In all there were 55 items.

Observational/Process Measures

Data on classroom behaviors were collected through the use of three different observational instruments that focused on (a) the teacher and aide, (b) the student, and (c) the classroom as a whole. Each is briefly described here.

Teacher-Aide Behaviors.   Periodic classroom observations were made through the use of an observation system designed to assess the frequency of the following behaviors:

1. Asking and/or responding to student questions
2. Facilitating performance on the activities (showing students how to complete the activity.)
3. Providing feedback as to the correctness of worksheet and/or activities.
4. Disciplining unacceptable behavior or providing management directions.
5. Providing direct instruction (see Berliner & Rosenshine, 1976).

Student Behaviors.   Student behaviors were observed periodically throughout the 14 weeks the study was in operation. Because of the difficulty and expense of obtaining these data, observations were made on only a subsample of students (N = 106). The subsample of "target" students was drawn on the basis of "relative language proficiency" according to the partitions reported in Duncan and De Avila (1979). In this way the subsample was composed of students who were proficient monolinguals in English or Spanish, fully proficient bilinguals, partially proficient in one or the other language, somewhat limited in both, or totally limited in both. See De Avila (1981) for a more detailed discussion of the procedures. In all verbal interactions, the extent of both English and Spanish usage was recorded. The specific behaviors observed are listed as follows:

1. Task-related talk
2. Requests for assistance
3. Offers of assistance
4. Non-task-related talk
5. Talk to teacher as opposed to others
6. Cleaning up
7. Working alone as opposed to working in either small or large groups
8. Observing others working
9. Waiting for directions from an adult
10. Task related transition versus "wandering"

Stated briefly, the purpose of the target child observation was to provide qualitative information as to language usage, socially defined learning behaviors, task-related interactions, and levels of engagement. The collection of these data enabled us to link classroom social behavior to learning outcome as a function of student linguistic characteristics. A more detailed discussion of the relative reliability and validity of the observational measures may be found in Cohen and Intili (1982) and in Cohen, Intili, and De Avila (1982).

Whole Class Behaviors.   To describe the organizational features of the classroom, grouping practices were observed. Weekly observations were made of each class as to size of groups, nature of activity, and interaction characteristics. Whole class observation was used also as a check for target student and teacher observations. For a more detailed description of the procedure and its agreement with individual observations see Cohen and Intili (1982).

Performance Measures

In addition to the data provided by paper-and-pencil and observational measures, data were also collected on the linguistic proficiencies of all participating students in both English and Spanish. Performance data on worksheets were collected also for target students.

Language Proficiency.   Proficiency in both English and Spanish was assessed in the beginning of the study and upon its completion by means of the Language Assessment Scales, Form A-Level I. (De Avila & Duncan, 1978). See De Avila and Duncan (1983) for a more detailed discussion of the psychometric properties of the test. The LAS is made up of the following subscales:

1. Phonemes
2. Minimal pairs
3. Vocabulary
4. Sentence comprehension
5. Production (story retelling)

Scores on the combined subtests are summed to provide a composite weighted score that represents the student's level of oral linguistic proficiency. Score values range from "no proficiency" to "fully proficient." The scores on the story retelling section of the test are the only subjective ratings; therefore, this section was blind-scored by non-project staff. Scoring for all of the other project-developed measures was conducted by project staff. Scoring of the CTBS was by the publisher.

Student Worksheets.   Worksheets were collected for each of these measures, and were scored for the following:

1. Total number completed
2. Accuracy of computations and/or descriptions
3. Use and quality of written language
4. Use and completeness of drawings
5. Complexity of reasoning and use of inference

Worksheets were available to students in both Spanish and English versions. The interrater reliabilities for the coding categories used to score worksheet performance are reported in Cohen and Intili (1982).

RESULTS

The results discussed were generated in several separate series of analyses. In the first series the general question of program effectiveness was addressed. In the second series, outcomes, student characteristics, and individual student classroom behaviors were examined in conjunction with classroom processes and instructional methods. Thus, the first series was directed at examining outcomes whereas the second was aimed at examining underlying processes. The analysis of outcomes was based on the entire sample whereas the analysis of process (observational) data was based on the target student subgroup.

Series I: Outcomes

Pre-post Gains. Treatment effects were examined by time, condition, school, sex and age for each set of dependent variables. Consistent statistically significant test gains were found on prepost comparisons for all major variables, i.e., CTBS, MINI, LAS, and so forth. Gains interacted with treatment and school variations. School variation, however, was associated with level of implementation (see Anthony et al., 1981). That is, although the FO/D period was intended to be 1 hour per day, some teachers allocated less time. Treatment was significantly related to five of the eight outcome measures. Examination of sex differences revealed that there were no meaningful differences between boys and girls as to either initial (pre-test) base rates or gains over time.

Age differences were also examined. In this series of analyses, significant age effects were found, suggesting that the program was least "effective" for a subgroup of older students. Examination of language data indicated that this group was made up of 14 monolingual Spanish speakers. Although there were other Spanish monolinguals in the sample, these 14 students tended to be substantially older than other students; in some cases by as much as 2 to 3 years. In subsequent discussion, it was suspected that since these students were recent immigrants, they might also occupy "low status" positions in the classroom (see Cohen & Intili, 1982). This would imply fewer verbal interactions with English-speaking peers and adults. The general importance of "talking and working together" is addressed in a later section.

Comparisons with Norms and Post Hoc Groups.   In a second phase of this series of analyses, the performance of participating students was compared to publisher norms and to post hoc comparison groups. Due to a number of problems beyond our control these comparisons were available for CTBS scores only. Consistent significant gains were found for both FO/D and comparison groups, which slightly favored the FO/D group. In a second set of comparisons, expected gains in the academic content areas (i.e., Reading and Math) were calculated by obtaining the difference between publisher raw-score norms equivalent to the 50th percentile for fall and spring test administrations. The differences between these two raw scores were then treated as "expected gain" scores and compared to pre/post gains scores obtained by FO/D and comparison students.

Results for the comparison group revealed that actual gains did not always match the Title VII program expectations. In several instances, as predicted from our initial content analyses of the CTBS subscales, comparison students fell further behind publisher norms on subscales requiring conceptual problem solving even though they improved in an absolute sense. When relative gains of the comparison students were compared to gains experienced by the FO/D group, significant differences were found in favor of the FO/D group on the predicted CTBS subscales, math application and concepts. Thus, although the comparison group was making substantial gains in math computation, a skill usually taught through rote methods, students were actually falling further behind the norms in the more conceptual skill areas.

Student Characteristics.   In an attempt to determine the extent of improvement as a function of student characteristics, three subanalyses were conducted. In the first, prior to the implementation of FO/D, a subgroup of "problem" (see Rosenholtz, 1982) students was identified (N = 60) from the 106 target students in the following manner. During the pre-test period, each teacher was asked to identify 6 to 8 students who they felt would be likely to experience difficulty in mastering the basic school curriculum and/or the science/math concepts embodied in the FO/D activities. By this procedure we were able to identify a group of students for whom the teacher held lower expectations. Comparisons between low expectation and remaining FO/D students showed low-expectation students to perform consistently lower at both pre- and post-test administrations. Nevertheless, the absolute gains for the two groups were virtually identical. In other words, even though the "problem" group scored lower at both points in time, their rate of improvement was indistinguishable from the rest of the group despite the lower "expectations."

In a second attempt to examine gain as a function of student characteristics, a subgroup was formed on the basis of intellectual development. For this purpose the CCS provided a means for identifying potentially "gifted" students. An arbitrary criterion score of 28 points out of a possible 32 was used to generate a group of 40 students. For a more complete discussion of the use of the CCS as a device for identifying "exceptional students" see De Avila (1980). Pre/post analyses were subsequently carried out on the outcome measures. Results revealed significantly greater gains across time for gifted students on most of the measures where comparisons were possible. The only exception to this finding was on the CTBS math computation subscale where improvement was substantially the same as for the rest of the sample. Gains in reading (Vocabulary and Comprehension) and math (Concepts and Applications) were of particular significance.

In a third and final attempt to examine program effects as a function of student characteristics, data were examined on the basis of linguistic considerations. This question was examined in two ways. In the first, the total sample was subdivided into two groups on the basis of oral English language proficiency scores (LAS). The resultant two groups were defined as "limited English proficient" (LEP) and "fluent English proficient" (FEP). A comparison of the relative pre/post gains for the two groups revealed significant improvement for both groups on most outcome measures. Further analyses revealed slightly stronger gains for the FEP group on the more traditional measures of school achievement, which included reading and math total scores and several subscales. On the other hand, LEP students showed slightly stronger gains on the CCS, LAS, and MINI tests.

The fact that the CTBS was administered only in English would account for the slightly stronger gains on the part of the fluent English speakers. Nevertheless, LEP students did show significant improvements on the CTBS. This improvement, however, was simply not quite as large as that of the fluent speakers. Furthermore, the importance of the linguistic factor is revealed by the fact that on the tests that students were able to take in either langauge, gains were stronger for the LEP group. Finally, the failure to find gains on the LAS for the FEP was certainly attributed to a ceiling effect and to the arbitrary definition of groups.

The analysis of data by LEP and FEP subgroups was conducted without consideration of home language proficiency. In order to gain insight into the nature of the interaction of the two languages, students were regrouped according to "relative linguistic proficiency." Using LAS scores, five linguistic subgroups were generated. They included:

1. English monolinguals.
2. Spanish monolinguals.
3. Limited speakers of both.
4. Minimal speakers of both.
5. Bilinguals (i.e., fully proficient in both).

Analyses of program effects by linguistic group showed that the proficient bilingual speakers exhibited the strongest and most consistent gains. Bilinguals were followed by limited bilinguals (Group 3) and Spanish monolinguals (Group 2) who were followed by minimal speakers (Group 4) and English monolinguals (Group 1). It is significant to note that even though there were between-group variations, all groups showed significant improvement on all measures except on the LAS, which was due to an artifact of group definition (ceiling effects).

Series II: Process Analyses

In an attempt to examine the relationship of "process" variables to learning outcomes, a number of analyses were conducted using "observational data" in conjunction with the outcome measures described above. Two sets of analyses are particularly important. In both instances, data were based on the target students. In the first set, Uyemura-Stevenson (1982) examined the relationship of student-student consultation to academic performance. In the second set, De Avila and Cohen (1983) compared teacher behavior in regular math classrooms with behavior observed during the FO/D class period, and then tied "method of instruction" to specific learning outcomes.

Uyemura-Stevenson (1982) examined three specific questions derived from sociological and psychological principles. In the first, the effect of student-student consultation was examined. Student-student consultation was defined as "task-related talk between students or other means of exchanging information about the task (i.e., modeling or demonstration)." Student consultation was operationalized on the basis of the observed behaviors coded on the "target student observation form" described above. Specific behaviors coded included task-related talk, requests for assistance from another student, offers of assistance, and working together without verbal interaction. Examination of the correlations between "talking and working together" and the outcome measures described above revealed significant relationships between student-student consultation and math conceptualization as measured on the student worksheet and total math scores as measured on the CTBS (see also, Cohen & Intili, 1982). In fact, it was found that "consulting with fellow students was a more powerful predictor of learning outcomes than consulting with teachers."

In addition, Uyemura-Stevenson found that student interaction facilitated performance on other academic behaviors such as the student's level of accuracy and use of written language (as opposed to drawing pictures) to describe the "procedures" and "results" of the activities. Thus Uyemura-Stevenson found that the extent of student interaction seemed to be related to the improvement of a fairly broad range of behaviors. It is important to bear in mind that the student data on which these analyses were conducted were based on a linguistically heterogeneous sample of students including monolinguals, proficient bilinguals, and children who had little proficiency in either standard Spanish or English.

One interpretation to these findings suggests that the linguistic variability of the sample may have actually contributed to gains in conceptual areas. Additional discussions of this speculation is made in a later section.

In a second set of analyses, Uyemura-Stevenson examined the extent to which consultation was more effective for students who exhibited "low academic and problem solving skills" and for those with higher skills in these areas. To some extent this question was similar to that addressed by De Avila and Cohen (1983), and by Rosenholtz (1981). Recall that Rosenholtz found that students who were seen by the teachers as having a low probability for success showed about the same improvement as other students although at lower absolute levels.

In order to test the notion of a differential effect, regressions were conducted for students divided at the median on prior academic or "cognitive resources." With respect to concrete tasks such as filling out the worksheets, analyses showed that student consultation was more important for students with fewer academic resources (i.e., lower prior academic achievement) than for students who had experienced less difficulty. On the other hand, results showed that student consultation was more effective for high achievers on more conceptual material.

However, Uyemura-Stevenson points out that there was an important relationship between "status" and student interaction that would have an impact on the extent of consultation between high- and low-status children. In a related study on these data, Cohen and Anthony (1982) found that "high status" students were more likely to be found "talking and working together" than children of lower status characteristics. It would appear that social status affects student interaction and that student interaction in turn affects learning outcomes. If status characteristics include teachers and peer perception of the child's ability to "speak English" then it would also seem clear that the problem of language difference faced by these students is exacerbated by a "social distance" that denies them access to verbal interaction.

In a third set of analyses Uyemura-Stevenson tested the proposition that combined teacher and student consultation would have an "interactive effect" that would lead to "better quality learning outcomes than would be expected from adding the effects of each type of consultation" (p. 90). When this proposition was tested, regression analyses failed to reveal any statistically significant "interaction effects" between teacher and student consultation; student-student consultation remained the most significant predictor of worksheet performance. According to Uyemura-Stevenson, the results of this set of analyses suggest that talking and working together seems to have two specific functions: (a) regardless of linguistic "deficiencies," it allows students to use one another as resources, and (b) it promotes interaction that in turn facilitates conceptual learning, particularly when the teacher is a scarce resource and unable to provide the type of "immediate feedback" widely advocated by such writers as Berliner and Rosenshine (1978).

De Avila and Cohen (1983) conducted secondary analyses of the data described above in which the relative effectiveness of "direct" and "indirect" instruction techniques were compared. In this series of analyses, De Avila and Cohen (1983) observed teacher and student behaviors in two settings and were able to link these behaviors to specific predicted academic outcomes.

The data consisted of observations made during FO/D and regular math periods. Original data collected by Rosenholtz (1981) had shown that the math classes and FO/D classes were two very different types of environments. Math classes were typified by "direct instruction" with the teacher and aide providing direct supervision over two or three "ability groups." In sharp contrast to the FO/D period, there was very little in the way of "student-student consultation" or verbal interaction between students. Rosenholtz's data also showed that engagement was higher under complexity.

As predicted, observed rate of "direct instruction" was related to gains on CTBS Math Computation, whereas rates of "talking and working together" were associated with gains in CTBA Math Applications. Their findings led De Avila and Cohen (1983) to conclude that there are interrelations between the nature of learning, type of classroom organization, and method of instruction that must be taken into account. Whole class instruction seems to be perfectly adequate for tasks requiring memorization of facts and figures. Conceptual learning, on the other hand, seems to imply a more complex classroom organization. In other words, given the complexity of the situation brought on by the diversity of student population and, by the nature of the task, the teacher has little choice but to "delegate authority" (see Cohen & Intili, 1982) when educational objectives require conceptual learning.

Nieves-Squires (1983) examined academic and linguistic gain as a function of relative linguistic proficiency and interaction. Results of this series of analyses revealed that, with slight variation, interaction covaried with academic and linguistic gain as a function of linguistic subgroup. Bilinguals were found to have had more total interaction than any other group with the exception of Spanish monolinguals. The verbal interaction of Spanish monolinguals, however, was probably restricted to itself by virtue of the group's monolingualism. Moreover, levels of interaction were found for the limited group with the other groups falling intermediate to the extremes.

CONCLUDING REMARKS

We have covered a wide number of issues of theoretical and applied interest. On the theoretical side we have reviewed research on the relationship between bilingualism and intellectual functioning. On the applied side we have considered a number of psychosocial issues as contributors to the school success of language minority students. Ostensibly, the first issue addresses the question of cognition and bilingualism whereas the second addresses the question of appropriate treatment for bilinguals. De Avila (1984) has recently suggested that the school behavior of language minority students could be understood as a function of three interacting factors; (a) intelligence/style, (b) interest/motivation, and (c) opportunity/access. In the following concluding remarks, rather than simply restating the results described in the above, we would like to discuss our results within the framework provided by these three factors.

Let us first consider the question of intelligence, cognitive style, and bilingualism as contributors to language minority performance on academic subject matter. Recall that for many, poor academic performance is attributable to, if not caused by, bilingualism. This conclusion seems faulty on three grounds.

First, in the review of various studies, it was found that conflicting results are the result of a failure to distinguish between ethnolinguistic group membership and relative linguistic proficiency. That is, the failure to control for the absolute language proficiency of comparison groups has resulted in a confounding of language with intellectual development and cognitive style.

Second, in past research where language proficiency differences have been controlled, few if any meaningful differences in intellectual development or cognitive style have been found that could be attributed to bilingualism. In this connection Saarni (1973), using Piaget-inspired tasks similar to those used in the present research, found that differences in cognitive style between boys and girls did not predict differences in intellectual functioning. Similar to our own work, Hyde (1981) reviewed the literature on sex differences in mathematical reasoning, conducted a meta-analysis of the results, and concluded that reported sex differences, although statistically significant, tended not to account for meaningful proportions of the total variance. Finally, we have found linguistic proficiency among language minority students to be a far stronger predictor of academic performance than either cognitive style or intellectual development. However, linguistic proficiency in English, although necessary, does not seem to be a sufficient condition.

Third, we found that when comparisons between the total group of FO/D students and norms on the CCS collected over the past 15 years (N = 6,000) were made, they failed to reveal any statistically significant differences. This group of students was no different from any other group with respect to intellectual development as measured by Piagetian tasks. On the other hand, the group was substantially behind their mainstream counterparts in level of academic performance. Although they seem to possess necessary levels of intellectual development, this, like linguistic proficiency per se, is not sufficient for school achievement.

Our conclusion on the question of the relationship between bilingualism and cognitive style or intellectual functions is threefold: (a) To the extent that differences in academic performance exist between language minority students, these differences cannot be attributed to the groups presumed bilingual. (b)

Students, to the extent they are proficient bilinguals, experience a wide range of cognitive and social advantages over other students. (c) Although linguistic proficiency and intellectual development are necessary conditions for the success of language minority students (and other children), they are not sufficient given traditional classroom practices and organization.

The second major factor in our framework deals with the issue of motivation and interest: the literature on differences between language minority and majority populations suffers from the same problems as the literature on intellectual and cognitive style differences. Although this chapter did not cover the issue of motivation in any direct sense, a few comments seem to be in order.

A number of recent writers have found student engagement to be a strong predictor of success in academic subject areas and have reported engagement rates on academic tasks to be as low as 10% of allocated time. Based on the finding that individualized seatwork tends to produce low levels of student engagement, Berliner and Rosenshine (1976) have concluded that the best way to remediate poor academic performance is by the means of "direct instruction." The major thrust of this methodological approach is to improve academic performance by increasing "time on task" through direct teacher supervision and whole class instruction. However, student engagement can be viewed as an index of interest and motivation to complete academic tasks. In this connection, Rosenholtz (1981) showed that direct instruction did not always maximize time on task, especially when it occurred in the complex multiple groups and classroom organizational structures such as those required by FO/D. Rosenholtz found that peer task talk was related to engagement in both math and FO/D classes. However, task engagement was correlated with direct instruction only in the math class and was uncorrelated in the FO/D period where there were multiple groups and activities in simultaneous operation. In general, it was found that the level of engagement during the FO/D period was in excess of 80%. What can be concluded from these data is that students were internally motivated to complete tasks they found interesting. We can further speculate that part of the interest derived from the peer interaction required to complete the tasks and not from the direct supervision of the teacher. Finally it would appear that different educational objectives require different instructional methods that, in turn, imply different classroom organizations.

The third factor in the facilitation of academic success deals with the issues of opportunity and access. Opportunity and access may be discussed at several levels including social, educational, and interpersonal. The concept of educational equity that motivates the justification, development, and funding of categorical programs, such as Title VII Bilingual Education, is based on the notion of "equal opportunity regardless of sex, religion, race or national origin." Although the avowed purpose of compensatory education programs has been to provide equal opportunity through remedial programs, it is doubtful that much of the social science research underlying the design of many of the programs actually provides equal access to education and intellectual development on a broader scale. In other words, although opportunity implies access it does not guarantee it. For example, the belief that English language and basic skill deficiencies preclude thinking and scientific reasoning has led to the establishment of programs for language minorities which, by default, emphasize rote memory skills at the expense of higher order intellectual processes. Thus, although students are provided with the opportunity for success by placement in special classes, presumed deficiencies, ironically, preclude access.

The social sciences have identified a wide variety of student characteristics thought to contribute to academic failure. What we have found is that the wholesale application of many of these findings is fraught with danger. Of particular importance to this discussion is the relation between social status variables operating both within and without the school setting and the extent to which differences (often more presumed than real) on these variables lead to differences in the design of programs that, in turn, preclude important process in learning. The case of student interaction is a prime example. What the previous findings have illustrated is that student interaction is a significant contributor to the learning process, particularly as related to concept formation. Interaction within the classroom, however, is to some extent modified by social status factors. To the extent that English language proficiency operates like other academic status variables, such as reading, it can be expected that students with limited proficiency will exhibit lower levels of interaction with English-speaking classmates. Finally, these data show that under classroom organizational conditions where language minority students are provided with access to multiple resources, including home language, peer consultation, manipulation and so on, they will acquire concepts as easily as mainstream students while at the same time acquiring English language proficiency and the basic skills. In fact, what the data show is that "bilinguals" are at a head start in this regard.

ACKNOWLEDGMENTS

The research reported herein was supported directly and indirectly by a variety of agencies including the National Institute of Education, the National Science Foundation, the California State Department of Education, the Ford Foundation, and Stanford University. I am particularly indebted to my colleagues, Elizabeth Cohen and Joann Intili who provided sociological insight, Sharon Duncan who provided linguistic understanding, and Cecilia Naverrette who assembled the task activities and has provided coordination at the classroom level over the past eight years. I would also like to thank Olivia Martinez and Auoura Quevedo of the San Jose schools who provided a place where we could test our ideas. I would like to thank Barbara Anthony, Sue Hanson, Susana Mata, Charles Parchment, Bruce Wilson, Nancy Stone, Brenda Uyermura-Stevenson, Stephen Rosenholtz, and Andrea Nieves, who have contributed in their special ways. Finally, I would like to thank Barbara Havassy who critically reviewed the work along the way.

An earlier version of this chapter was presented at a conference on childhood bilingualism sponsored by the Society for Research in Child Development that was held in New York during the summer of 1982.

REFERENCES

Anthony, B., Cohen, E. G., Hanson, S. G., Intili, J. K., Mata, S., Parchment, C. Stevenson, B., & Stone, N. (1981, April). *The measurement of implementation: A problem of conceptualization.* Paper presented at the annual meeting of the American Educational Research Association, Session 40.11.

Arias, B. (1982, May). *Contextual variation and the implementation of the bilingual curriculum.* Paper presented at the annual meeting of the American Educational Research Association, New York.

Bellack, A. A., Kliebard, H. M., Hyman, R. T., & Smith, F. L. (1966). *The language of the classroom.* New York: Teachers College Press.

Berliner, D., & Rosenshine, B. (1976). The acquisition of knowledge in the classroom. In R. C. Anderson, R. J. Spiro, & W. E. Montague (Eds.), *Schooling and the education process* (pp. 375–98). Hillsdale, NJ: Lawrence Erlbaum Associates.

Bourne, L. E. (1966). *Human conceptual learning.* Boston: Allyn & Bacon.

Bruner, J. S., Oliver, R., & Greenfield, P. M. (1966). *Studies in cognitive growth.* New York: Wiley.

Buriel, J. W. (1975). Cognitive styles among three generations of Mexican-American children. *Journal of Cross-Cultural Psychology, 6,* 417–39.

Chan, K. S. (1983). Limited English speaking, handicapped, and poor: Triple threat in childhood. In M. Chu-Chang (Ed.), *Asian and Pacific American perspectives in bilingual education: Comparative research* (pp. 153–168). New York: Teachers College Press.

Clark, G. M., & Peterson, P. L. (1976, April). *Teacher-stimulated recall of interactive decisions.* Paper presented at the annual meeting of the American Education Research Association, San Francisco, CA.

Cohen, E. G., & Anthony, B. A. (1982, March). *Expectation states theory and classroom learning.* Paper presented at the annual meeting of the American Educational Research Association, New York.

Cohen, E. G., Deal, T. E., Meyer, J. W., & Scott, W. R. (1976). *Organization and instruction in elementary schools* (Technical Report No. 50). Stanford, CA: Stanford Center for Research and Development in Teaching.

Cohen, E. G., & Intili, J. K. (1982). *Interdependence and management of bilingual classrooms* (Final Report to NIE). Stanford, CA: Stanford University, School of Education.

Cohen, E. G., Intili, J., & De Avila, E. A. (1982). *Learning science in bilingual classrooms: Interaction and social status* (Final Report, National Science Foundation Grant #SED 80-14079). Stanford, CA: Stanford Unviersity, School of Education.

De Avila, E. A. (1980). *The cartoon conservation scales: Implications for the classroom - A NeoPiagetian approach.* San Rafael, CA: Linguametrics Group.

De Avila, E. A. (1981). *Improving cognition: A multicultural approach* (Final Report, National Institute of Education Grant NIE-G-78-0158). Stanford, CA: Stanford University, School of Education.

De Avila, E. A. (1984). Motivation, intelligence, and access: A theoretical framework for the education of minority language students. In *Issues in English language development* (pp. 21–31). Rosslyn, VA: National Clearinghouse for Bilingual Education.

De Avila, E. A., & Cohen, E. G. (1983). *Indirect instruction and conceptual learning* (Report to NIE). Stanford, CA: Stanford University, School of Education.

De Avila, E. A., & Duncan, S. E. (1978). *Language assessment scales (LAS)*. Corte Madera, CA: Linguametrics Group.

De Avila, E. A., & Duncan, S. E. (1979). Bilingualism and metaset. *NABE Journal, 3*(2), 1–20.

De Avila, E. A., & Duncan, S. E. (1980). Definition and measurement of bilingual students. In *Bilingual program, policy, and assessment issues* (pp. 47–62). Sacramento, CA: California State Department of Education.

De Avila, E. A., & Duncan, S. E. (1982). *Finding out/Descubrimiento*. San Rafael, CA: Linguametrics Group.

De Avila, E. A., & Duncan, S. E. (1983). *The language minority child: A psychological, linguistic, and educational analysis*. San Rafael, CA: Linguametrics Group.

De Avila, E. A., Duncan, S. E., & Ulibarri, D. M. (1982). Cognitive development. In E. E. Garcia (Ed.), *The Mexican American child: Language, cognition, and social development* (pp. 59–106). Tempe, AZ: Center for Bilingual/Bicultural Education.

De Avila, E. A., Duncan, S. E., Ulibarri, D. M., & Fleming, J. S. (1979). *Predicting the academic success of language minority students from developmental, cognitive style, linguistic and teacher perception measures.* Study conducted as part of a three-year cross-cultural investigation of cognitive styles of eight ethno-linguistic groups, supported by Contract #400-65-0051 between the National Institute of Education and the Southwest Educational Development Laboratory.

De Avila, E. A., & Havassy, B. E. (1975). Piagetian alternatives to IQ: Mexican American study. In N. Hobbas (Ed.), *Issues in the classification of exceptional children* (pp. 246–266). San Francisco: Jossey Bass.

De Avila, E. A., Havassy, B. E., & Pascual-Leone, J. (1976). *Mexican-American school children: A neo-Piagetian analysis*. Washington, DC: Georgetown University Press.

De Avila, E. A., & Pulos, S. M. (1978). Developmental assessment by pictorially presented Piagetian material: The cartoon conservation scales (CCS). In G. I. Lubin, M. K. Poulsen, J. F. Magary, & M. Soto-McAlister (Eds.), *Piagetian theory and its implications for the helping professions* (pp. 124–39). Los Angeles, CA: University of Southern California.

Duncan, S. E. (1979). *Child bilingualism and cognitive functioning: A study of four Hispanic groups*. Unpublished doctoral dissertation, Union Graduate School.

Duncan, S. E., & De Avila, E. A. (1979, February). *Relative linguistic proficiency and field dependence/independence: Some findings on the linguistic heterogeneity and cognitive style of bilingual children.* Paper presented at the 13th Annual Convention of TESOL, Boston, MA.

Fleming, J. S., & De Avila, E. A. (1980). Scalogram and factor analyses of two tests of cognitive development. *Multivariate Behavioral Research, 15,* 73–93.

Goodenough, D. R., & Karp, S. A. (1961). Field dependence and intellectual functioning. *Journal of Abnormal Psychology, 63,* 241–246.

Hansen, S. (1980). *Psychometric properties of the MICA mini-test.* Unpublished manuscript, Stanford University.

Hunt, J. M. (1961). *Intelligence and experience.* New York: Ronald Press.

Hyde, J. (1981). How large are cognitive gender differences? *American Psychologist, 36*(8), 892–901.

Intili, J. K. (1977). *Structural conditions in the school that facilitate reflective decision making.* Unpublished doctoral dissertation, School of Education, Stanford University.

Intili, J. K., & Flood, J. E. (1976, May). *The effect of selected student characteristics on differentiation on instruction in reading.* Paper presented at the annual meeting of the American Educational Research Association, San Francisco.

Laosa, L. M. (1977). Cognitive styles and learning strategies research: Some of the areas in which psychology can contribute to personalized instruction in multicultural education. *Journal of Teachers Education, 28*(3), 26–30.

Lochhead, J. (1985). Teaching analytic reasoning skills through pair problem solving. In J. Segal & S. Chipman (Eds.), *Thinking and learning skills: Vol. 1. Relating instruction to research* (pp. 1–19). Hillsdale, NJ: Lawrence Erlbaum Associates.

Mestre, J. P., Gerace, W. J., & Lochhead, J. (1982). The interdependence of language and translational math skills among bilingual Hispanic engineering students. *Journal of Research in Science Teaching, 19,* 399–410.

Nieves-Squires, H. A. (1983). *Talking in the classroom and second language acquisition.* Unpublished doctoral dissertation, Stanford University.

Perrow, C. (1967). A framework for the comparative analysis of organizations. *American Sociological Review, 32,* 194–208.

Ramirez M., III (1973). Cognitive styles and cultural democracy in education. *Social Science Quarterly, 53*(4), 895–904.

Riessman, F. (1962). *The culturally deprived child.* New York: Harper & Row.

Rosenholtz, S. (1981). *Effects of task arrangements and management systems of task engagement of low achievement students.* Unpublished doctoral dissertation, Stanford University.

Rosenthal, A. S., Milne, A., Ginsburg, A., & Baker, K. (1981). *A comparison of the effects of language background and socioeconomic status on achievement among elementary school students.* Washington, DC: AUI Policy Research, Office of Planning and Budget, Department of Education.

Saarni, C. I. (1973). Piagetian operations and field independence as factors in children's problem-solving performance. *Child Development, 44,* 338–345.

Sanders, M., Scholtz, J. P., & Kagan, S. (1976). Three social motives and field independence-dependence in Anglo American and Mexican American children. *Journal of Cross-Cultural Psychology, 7*(4), 451–462.

Shavelson, R. J., & Stern, P. (1980). Research on teachers' pedagogical thoughts, judgements, decisions, and behavior. *Review of Educational Research.*

Sternberg, R. J. (1981). Testing and cognitive psychology. *American Psychologist, 36*(10), 1181–1189.

Taba, H. (1966). *Teaching strategies and cognitive functioning in elementary school children* (U.S.O.E. Cognitive Research Project No. 2404). San Francisco: San Francisco State College.

Takahashi, Y. (1982). *The effects of modified input and interaction patterns on language acquisition.* Unpublished manuscript, Stanford University.

Uyemura-Stevenson, B. (1982). *An analysis of the relationship of student-student consultation to academic performance in differentiated classroom settings.* Unpublished doctoral dissertation, Stanford University.

Valett, R. E. (1978). *Developing cognitive abilities: Teaching children to think.* St. Louis: C. V. Mosby.

Wilson, B., De Avila, E. A., & Intili, J. K. (1982, May). *Improving cognitive, linguistic, and academic skills in bilingual classrooms.* Paper presented at the annual meeting of the American Educational Research Association, New York.

Chapter 8

The Mathematics Achievement
Characteristics of Asian-American Students

Sau-Lim Tsang
ARC Associates, Inc.,
Oakland, CA

The Asian-American population of the United States has experienced a dramatic growth since 1970 because of immigration and refugee resettlement (Tsang & Wing, 1985). The 1980 Census (Bureau of the Census, 1981) counted 3.5 million Asian-Americans. Others (e.g., the Center for Continuing Study of the California Economy) have estimated the 1985 Asian-American population at 5.5 million. Coincident with this rapid increase is the diversification of the population, which consists of Cambodian, Chinese, Filipino, Hmong, Indian, Japanese, Korean, Laotian, Mien, Vietnamese, and many others. This increase and diversification have created much visibility and interests in the Asian-American population. One of the most sensational media topics is the educational achievement of Asian-American students, especially in mathematics and related subject areas.

Despite the perceived success, fostered in large part by the mass media and supported by anecdotal evidence, few researchers have systematically examined the mathematics achievement of Asian-American students. Two factors may have contributed to this lack of interest. First, Asian-Americans have been considered a numerically insignificant demographic group, and research and policy agendas have given priority to other students. Second, the well-publicized high mathematics achievement of Asian-American students fostered the assumption that Asian-American students have few mathematics learning problems and therefore no research is needed.

However, the lack of a comprehensive body of knowledge about the mathematics education of Asian-American students tends to lead to inappropriate action (or inaction) concerning the schooling of Asian-Americans on the part of

123

policy makers and educational practitioners. Further, failure to understand the mathematical achievement of Asian-American students is a failure to tap information that may benefit our understanding of American education and how it might be changed for the betterment of all students. Somewhat ironically, business and educational leaders have looked instead to the schooling of students in Asian countries (e.g., China, Japan, and Taiwan) for sources of enlightenment in assessing the condition of the American educational system.

In this chapter, the author (a) surveys recent data sources to develop a mathematical achievement profile of Asian American students; (b) discusses three endemic factors leading to this profile; and (c) discusses the influence of the immigrant students' educational background on their mathematics learning styles, understanding of mathematics concepts, and test-taking abilities.

MATHEMATICS ACHIEVEMENT

Tsang (1984) surveyed the literature on the mathematics achievement of Asian-American students and concluded that Asian-American students were achieving at or slightly above the level of white students. However, the survey was based mostly on data sources collected before or immediately after 1965, just before the large number of Asian immigrants and refugees began to enter the United States and drastically change the composition of the Asian-American population. A review of more recent data is thus warranted to update the mathematical achievement profiles of Asian-American students. Two sources would provide us with the needed up-to-date data: the Scholastic Aptitude Test (SAT) results and the High School and Beyond study.

Since 1980, the College Entrance Examination Board has compiled and released annual reports on SAT scores and SAT candidates. The information about the candidates is gathered through a student descriptive questionnaire that is part of the SAT registration form. The SAT data show that in 1982–83, 33,062 out of the total of 983,474 candidates attending high school in the 50 states and Washington, DC, identified themselves as Asian-Americans. This is approximately 46% of all 17- to 18-year-old Asian-Americans. In comparison, the total number of SAT candidates is only 24% of the total 17- to 18-year-old population (Ramist & Arbeiter, 1984). Table 8.1 shows that the mean SAT math score of Asian-American candidates was substantially higher than that of the white candidates. However, there is a major shortcoming of the SAT scores of Asian-Americans. The College Entrance Examination Board included scores of foreign students with those obtained from Asian-Americans. In 1982-83, SAT candidates in foreign countries numbered 3,719 (10%) of the total of 36,781 candidates labeled Asian-Americans. This confounding of foreign candidate data probably inflated the mean SAT math scores of the Asian-Americans.

TABLE 8.1
Percentage Distribution of Asian and White SAT Math Scores
in 1982–83

White		Asian*
X	750–800	XXX
XXX	700–749	XXXXXX
XXXXX	650–699	XXXXXXXX
XXXXXXXXX	600–649	XXXXXXXXXXX
XXXXXXXXXXXX	550–599	XXXXXXXXXXXXX
XXXXXXXXXXXXXXXX	500–549	XXXXXXXXXXXXXX
XXXXXXXXXXXXXXXX	450–499	XXXXXXXXXXXXX
XXXXXXXXXXXXXXX	400–449	XXXXXXXXXXXX
XXXXXXXXXXXX	350–399	XXXXXXXXXX
XXXXXXXX	300–349	XXXXXXX
XXX	250–299	XXX
	200–249	X
Mean = 484 (S.D. = 114)		Mean = 514 (S.D. = 127)

*Includes 10% foreign students
Source: Ramist and Arbeiter (1984)

Table 8.2 provides an indication of the effects of foreign candidates on the SAT math scores. When the students were grouped according to their best language, those Asian-American candidates who said English was their best language (and probable non-foreign candidates) scored slightly lower than those who said English was not their best language. Both groups scored substantially higher than their white counterparts.

TABLE 8.2
1982–83 SAT Math Scores by English as Best Language

% of Total	SAT-M Percentile Score		
	25th	50th	75th
Asian*			
Yes 71.6	419	515	606
No 28.4	412	520	619
White			
Yes 98.2	400	481	565
No 1.8	357	443	533

*Includes 10% foreign students
Source: Ramist and Arbeiter (1984)

High School and Beyond is a longitudinal study being conducted by the U.S. Department of Education. The study sample consists of 30,000 students who were sophomores and 28,000 who were seniors during 1980, the base year for data collection. Questionnaires and a test battery were administered to the students; a follow-up survey of the same students was conducted in 1982, and additional follow-ups are planned. The High School and Beyond test data show that the average Asian-American student score in mathematics was slightly higher than the average white student score. When analyzed by length of residence in the United States, the data showed that those who have resided in the United States for 6–10 years scored especially higher than other Asian-American and white students (Table 8.3).

As part of the first High School and Beyond follow-up survey in 1982, the achievement tests were readministered to those who were sophomores in 1980. Table 8.4 shows an analysis of the improvement in test scores. All Asian-American students categorized by length of residence improved their scores by about the same amount as the white students. However, the 6-10-year group again outperformed other Asian-American and white groups.

The above two data sources and others reviewed by Tsang (1984) all indicate that Asian-American students are achieving at least as well as the white students in mathematics. However, one must be cautioned that the comparison is on mean scores of distributions. For Asian-Americans, the standard deviations of the

TABLE 8.3
Average Test Scores of Asian and White High School
and Beyond Cohorts in 1980

| API | 10th Grade | | 12th Grade | |
Time spent in US	Math (max = 38)	Science (max = 20)	Math (max = 38)	Science (max = 20)
All or almost all life	17.4	10.1	19.0	11.3
	(9.94)	(4.65)	(11.10)	(5.20)
More than 10 yrs.	18.1	10.1	17.1	11.6
but not all life	(10.94)	(4.92)	(12.52)	(4.28)
6–10 years	16.6	9.9	21.3	11.7
	(9.51)	(3.03)	(9.74)	(2.92)
1–5 years	15.6	6.6	18.7	9.0
	(11.27)	(4.53)	(12.22)	(4.86)
All APIs	16.8	9.2	19.1	10.8
	(10.33)	(4.71)	(11.32)	(4.89)
White	15.5	10.3	17.8	11.3
	(9.40)	(4.10)	(10.01)	(4.07)

Note: Standard deviations are in parenthesis; (max = number) refers to the maximum possible test scores.

Source: Peng et al. (1984) and personal communication with Peng.

TABLE 8.4
Average Percentage of Test Items Correctly Answered
by Sophomore Cohort Members in 1982
That Were Incorrectly Answered in 1980

	Mathematics		Science	
Asian				
All or almost all life in U.S.	45%	(23.0%)	41%	(21.3%)
More than 10 years but not all life	41%	(20.3%)	45%	(24.6%)
6–10 years	50%	(23.2%)	45%	(17.1%)
1–5 years	46%	(25.0%)	41%	(18.9%)
All Asian	45%	(23.5%)	42%	(20.4%)
Whites	44%	(20.8%)	40%	(21.1%)

Note: Standard deviations are in parentheses.
Sources: Peng et al. (1984) and personal communication with Peng.

distributions are uniformly larger than the other ethnolinguistic groups indicating a larger spread in the distribution of scores. In other words, although there is a higher proportion of Asian-American students who are high math achievers, there is also a larger proportion who are low math achievers. The mean scores only provide us with a generalization and do not reflect the diverse achievement levels of Asian-American students.

FACTORS AFFECTING MATH ACHIEVEMENT PROFILE

Immigration and Refugee Policy

The present immigration policy of the United States was legislated in 1965, when Congress enacted a bill that admitted 20,000 immigrants per country per year. The law, which took effect in 1968, reversed nearly 80 years of exclusion of Asian immigrants. Not only did the new policy have a major impact on the size and ethnic diversity of the Asian-American population, but it also influenced the educational characteristics of those admitted. Because the majority of Asian-Americans are immigrants, the educational characteristics of the overall population are similarly a function of the immigration policy.

The present immigration policy has the dual objectives of reunifying families and increasing the supply of needed labor (U.S. Commission on Civil Rights, 1980). Based on these objectives, wives and minor children of U.S. citizens are admitted on a nonquota basis, and the 20,000 slots allotted to each country are rationed between two groups: (a) other relatives of U.S. citizens and lawful resident aliens, and (b) professional and other workers needed by American employers. Thus, the first cohorts of Asian immigrants admitted under this system consisted of relatives of Asians who had come here much earlier, and skilled, highly educated workers for whom there were employment opportunities in this country. The first group, whose admittance was based on their relationship to relatives who came to this country primarily as unskilled, uneducated laborers from rural regions, probably came from lower socioeconomic backgrounds. The second group, mostly composed of people from math- and science-related professions, came from higher socioeconomic backgrounds. Over time, this dichotomy became less well defined as relatives of the professionals and other skilled workers began immigrating under the family reunification categories. In general, the socioeconomic background of most present-day Asian immigrants is likely to be middle class.

Contributing to this phenomenon are the Asian foreign students and Southeast Asian refugees. A large number of Asians come to the United States for higher education and do not return to their homelands. Foreign students are typically subject to much higher tuition and fees than American residents, and they must be able to afford expensive transportation costs to and from their home countries and American universities. Further, their opportunity to earn income while enrolled in American schools is severely limited by the federal government. Thus, it is likely that most of these students are from relatively high socioeconomic backgrounds.

A substantial proportion of recent Southeast Asian refugees might also be categorized as middle class. Many of the so-called first wave Southeast Asian refugees, those who were admitted to the United States between 1975 and 1979, were from the elite class of South Vietnam. Among them were former government officials and scholars trained at prestigious universities in France.

The high level of education and socioeconomic status of the first wave of refugees, foreign student immigrants, professional and other skilled workers and their relatives was likely to influence their children's education in two ways. First, the high socioeconomic status positively affected high educational achievement and in this instance, mathematical achievement. Second, because of the influence of their parental occupations, Asian-American students place emphasis on mathematics achievement, which is a prerequisite for college admission and subsequent professional occupations.

The immigration and refugee policy may also be linked to the large standard deviations of the test scores described earlier. To review, the large standard deviation indicates a wider distribution for the Asian-American population and thus a comparatively greater proportion of low achievers among Asian-Americans. These low achievers may be those from immigrant families with relatively low socioeconomic backgrounds. According to the 1980 Census (Bureau of the Census, 1984), 14% of Asian-Americans were living in poverty compared to 9% of the white population. Many so-called second wave Southeast Asian refugees, those admitted since 1979, come from rural, preliterate societies; it was estimated that 76% of this group of refugees were welfare dependent in 1982 (Ford Foundation, 1983). It is also likely that the influx of immigrants who are related to early Asian immigrants from rural areas has not yet terminated.

Finally, the high proportion of immigrants among the Asian-American population means that many children are not fully proficient in English. This non-English language background may have influenced them to concentrate on mathematics, resulting in higher scores on mathematics tests. This suggestion is borne out in part by a research study by Bagasao (1983), who found that Asian-American high school students with 5 or fewer years of residence in the United States were more likely to plan "science careers" than Asian-American students who had been in the country longer or who were American born (Table 8.5).

TABLE 8.5
Career Plans of High School and Beyond Asian Seniors

Length of Residence	Science Career (N = 102)		Non-science Career (N = 102)	
	%	(N)	%	(N)
U.S. born and raised	40.0	(46)	60.0	(69)
Foreign born				
Resident of U.S. 6 yrs. or	50.9	(20)	49.1	(15)
more	66.7	(36)	33.3	(18)
Resident of U.S. 5 yrs. or less				

Source: Bagasao (1983)

The High School and Beyond data also indicate that the group that has resided in the United States from 6 to 10 years is performing especially well when compared with other Asian-American and white groups. This high educational achievement may well be the result of the group's "better" educational preparation received in their home countries.

Time Spent on Learning

Asian-Americans appear to spend more time on learning than other students. Data from the High School and Beyond study show that, compared to white seniors, the Asian-American seniors took 1.5 more years of the "new basics," that is, academic subjects, and a higher percentage of the Asian-American sophomores spent 5 or more hours per week on homework (Table 8.6). Asian Americans were also less likely than other students to be absent from school (Peng, Jeffrey, & Fetters, 1984). Although 26% of white sophomores were never absent, 45% of Asian-American sophomores had perfect attendance. SAT data also indicated, even among high school students who were all college-bound, that

TABLE 8.6
Time Spent on Learning by High School and Beyond Students

Racial/Ethnic Group	Credits Earned in All Subjects	Credits Earned in New Basics Only	Sophmores Spending 5 or More Hours Per Week on Homework
Asian	22.6	14.7	46%
White	21.9	13.2	29%
Black	21.1	11.9	25%
Hispanic	21.7	11.7	16%
American Indian	21.3	11.2	22%

Note: One credit is earned for a one-year course. The new basics are English, mathematics, science, social studies, foreign language and computer science.

Source: Peng et al. (1984)

Asian-Americans reported they had 16.8 years of academic study compared to 16.32 for all students (Ramist & Arbeiter, 1984). The extra time Asian-American high school students appear to devote to learning is probably related to their general academic achievement and, in particular, to their mathematical achievement. For example, Peng et al. (1984) found that the number of credits earned in high-level mathematics courses is the second best predictor for mathematics achievement (after previous mathematics achievement scores).

As to why Asian-American students spend more time on learning than other students, a study by Stevenson (1983) suggests one possibility. In his longitudinal, comparative study of students in Taiwan, Japan, and the United States, Stevenson asked the mothers if luck, ability, or effort was the critical factor underlying their children's academic performance. Most Asian mothers chose effort, whereas most American mothers selected ability. The belief that achievement depends more on effort than ability may be similarly prevalent among Asian-American parents and transmitted to their children. If so, the greater amount of time Asian-American students spend on learning may represent extra effort expressly for the purpose of doing well. However, emphasizing the importance of effort is not particular to Asian culture. One of the reputed bedrocks of American culture is the belief that hard work leads to success. Stevenson's finding about American mothers' perceptions of the role of innate ability warrants further investigation.

Historical Labor Market Discrimination and Asian-American Sensitivity to Job Openings Under Conditions of Equal Employment Opportunity

Early Asian immigrants were welcomed by employers when there were labor shortages. Early Asian immigrants in the United States worked primarily in low-level manual labor jobs in agriculture and in the incipient urban-based industries of the West. These could be called "immigrant jobs," that is, low paying and low status jobs that domestic workers shunned. Upward occupational mobility was difficult, and when the shortages disappeared or if an economic recession appeared imminent, the immigrants were the first to suffer layoffs and other negative consequences. Thus, the jobs that immigrants did secure tended to be of short-term duration and were dead-ends in terms of upward mobility. Over time, Asian-Americans seemed to have developed a particular strategy to deal with employment discrimination and to secure upward mobility. This strategy has had an impact on their educational profile.

Beginning with World War II, when there was an economic boom in war-related industries and the first federal equal employment opportunity policies were adopted, second generation Asian-Americans found new occupations open to them just as other minorities and women did. But because skilled, industrial, union jobs had long been closed to Asian-Americans, Asian-American youth sought employment in other sectors of the economy. There was a new, rising need for science- and engineering-trained workers and, perhaps because hiring appeared to be based on merit, Asian-Americans began to enter these professional and technical occupations. In order to qualify for these types of jobs, they invested in college education and focused their studies in mathematics and related subjects. Between 1940 and 1950, there was a threefold increase in the number of Chinese-American males employed as professional, technical, and kindred workers (Lee, 1960). Although 110,000 people of Japanese ancestry were imprisoned during World War II, early releases were permitted for college attendance in the Midwest and East; and according to Kitano (1969), Japanese-Americans were able to capitalize on professional job opportunities in the post-internment years.

These first entrants into professional and technical fields became role models for subsequent cohorts of Asian-Americans who exhibited the same sensitivity to opportunities in professional fields during the Sputnik era and the subsequent period of growth among high technology industries. Further, once the 1964 Civil Rights Act was passed, the more blatant forms of employment discrimination became illegal. Altogether, historical job discrimination, job market sensitivity, and equal employment opportunity policies appear to have encouraged Asian-Americans to do well in mathematics preparation for college and subsequent careers in professional occupations.

MATHEMATICS CHARACTERISTICS OF ASIAN IMMIGRANT STUDENTS

The large number of Asian-American students who were born and received part of their education in their native countries has significant implication for the design of mathematics curriculum and instruction. These students were influenced by the educational system of their native countries, which may have differed considerably from that of the United States. The following is a summary of the few studies that examined how their educational background affected their learning styles, understanding of mathematics concepts, and test-taking abilities.

Tsang (1983) conducted a case study of eight Chinese immigrant students from Vietnam, China, and Hong Kong. The students had been in the United States less than 2 years and were selected from a high school Algebra I class for Chinese limited-English-proficient students. The study found that a common characteristic of the mathematics curriculum taught to the eight students before they came to the United States was an emphasis on tests. All were from countries with very competitive educational systems. Every student had to pass a series of public examinations in order to gain entrance to higher levels of schooling. Thus, the main objective of the teachers was to help the students obtain high scores on the examinations. Mathematics was taught with the same objective; teachers concentrated on the students' ability to do problems quickly and accurately. The students, therefore, learned many rules and formulae to solve different types of problems expediently. Because they were accustomed to this type of orientation, the students were perplexed by the Algebra 1 curriculum and the instructor's emphasis on the understanding of concepts. In general, they said that the explanations of concepts were clearer, but the way they were presented made the concepts too profound and complicated. When the homework assignments could be done with the rules and formulae they had learned before immigrating, they would ignore the usually lengthier approach taught in the class and proceed with their own methods. Consequently, many forgot most of what was taught in the class. Further, except for one, none of the eight students had the habit of checking the results or solution. This may be the result of training in their home country where speed was emphasized so that they could do as many problems as possible within a time limit, a skill that was essential to obtaining high scores in examinations.

The study also found that there was much cooperative learning among the students. Encouraged by the relaxed style of the Algebra I teacher, the students interacted quite freely to assist each other during class work. Many of the rules and formulae, though not taught by their teacher, were transferred from one student to another. Having acquired these rules and formulae to facilitate their problem solving, the students would abandon the more cumbersome procedures learned in class.

The students also showed inexperience with problems that did not ask for a specific solution or a set of solutions, that is, problems that asked for proofs or explanations. The students had seldom encountered such algebraic problems in their home countries, and many were lost when faced with such tasks. They did not know how to write proofs or explanations.

In another study, Ng and Tsang (1980) compared the performance of three groups of students: Americans of European descent, Chinese-Americans, and Chinese in Hong Kong. A battery of ability tests was administered together with a word association test and a sorting test. The results of the two tests showed that for all three groups of students the concepts were clustered into three categories: geometry, set theory, and number theory. The students from Hong Kong resembled the two groups of American students in their conceptions of set theory and the four operations of arithmetic, but the American students made a closer association between geometric and rational-number concepts than the Hong Kong students. This study suggested that students educated in Hong Kong perceived mathematical concepts differently than students whose entire education had been in the United States. Further studies of specific differences in concept structure between students from various countries are needed, however, before one could apply this finding to classroom instruction and curriculum development.

Educators have also shown concern about the reliability and validity of standardized mathematics tests for Asian immigrant students because of the demand for English proficiency and the cultural relevance of the test items. Tsang (1976) studied the effects on Chinese students of the linguistic and cultural assumptions of standardized mathematics tests. Twenty word problems with contents unfamiliar to the Chinese immigrant students were selected from several such tests, and four versions of a test were constructed. One version contained the original problems in English, one gave the problems in both English and Chinese, one was in English with the problems modified to eliminate cultural biases, and one gave the modified problems in both English and Chinese. The four versions had high reliability coefficients. Recent immigrants—that is, students who had lived in the United States less than 3 years—scored higher on the bilingual versions than the English versions. When the immigrants were divided into high and low achievers, only the low achievers scored higher on the modified problems than on the original problems. For those students who had resided in the United States more than 3 years, there were no significant differences in the difficulty of the four versions. The immigrant students probably benefited from the bilingual versions of the test because of limited proficiency in English. The cultural bias of the mathematics achievement test, though controversial, did not appear to have much effect on the students' performance. The higher achievers in particular seemed able to overcome the biased content of the problem.

Tsang (1983) also found that English proficiency affected the Chinese immigrant students' understanding of word problems differently according to their mathematics abilities. For higher achievers, English did not seem to have much of an effect on comprehension. Somehow, they were able to decode the mathematical contents underlying all the English words. In fact, they were able to derive the meanings of many English words from their understanding of the mathematical relationships in the word problem. The lower achievers were more dependent on their English proficiency. They usually said that they did not understand the problem because there were one or more English words they did not know. But often, after the researcher had translated every word to them, they still could not understand the problem.

Educators have also studied the effectiveness of training minority students in test-taking strategies. Bernal (1972) and Ginther (1978) found that such training improved Chicano students' performance on tests with unusual formats or tests that required a nonstandard test-taking strategy. Chinese immigrant students seem to encounter difficulties with such tests. Tsang (1976) administered a mathematics word association test to a group of Chinese immigrant students. The students were to write down in one minute as many mathematical concepts as they could think of that were related to a stimulus word from mathematics. Over half the students failed to write any response. When the test was subsequently administered to a comparable group of students who had been trained in test-taking strategies, they had no difficulty completing the test (Ng & Tsang, 1980).

CONCLUSION

A salient caveat when examining the Asian-American population is its diversity. Asian-Americans came to the United States from many different cultural backgrounds, societal conditions, economic status, and for different motivations. Generalizations are often simplifications of complex phenomena or explanations and must be examined carefully when applying them to a specific population.

This chapter first reviewed data sources that reveal that Asian-American students are achieving at least as well as white students in mathematics mean test scores. However, the large standard deviations also indicate the diversity of the population and that there is a larger proportion of Asian mathematics low-achievers.

Next, the author proposed that immigration policy, time spent on learning, and historical labor market discrimination and Asian-American awareness of subsequent labor market opportunities under equal employment opportunity conditions were major influences on the mathematics achievement profiles of Asian-Americans. Again one must be cautioned that these three factors may not contribute to a compelling explanation of a particular Asian-American group's mathematical achievement profile. For example, current immigration policy does not have a large impact on the Japanese-American educational profile, and the influence of past job discrimination on ethnic groups such as the Vietnamese who have little history in the United States is unsure.

The two factors, time spent on learning and awareness of employment opportunities, are particularly important. Not only may they be valid for all ethnic groups of Asian-American students, but they are also factors that the educational system can influence with respect to students of other racial and ethnic backgrounds. Time spent on learning can be increased, particularly if business, education, and government verbalize and show by example the relationship between effort and achievement in school, work, community service, and other activities that contribute to the quality of life. Greater awareness of work options and the preparation needed for employment might be fostered through more counseling and guidance programs and closer relationships between schools and businesses.

This chapter examined a set of studies and summarized how the immigrant students' educational experience in their native country influenced their learning styles, understanding of mathematics concepts, and test-taking abilities. All of the studies reviewed were conducted on Chinese immigrant students and generalizations to other Asian groups must be made with extreme care. Nevertheless, findings suggest that mathematics instruction for Asian immigrant students should place special emphasis on the understanding of concepts and algorithms, one of the major objectives of the U.S. mathematics curriculum. On testing, the findings suggest that mathematics tests administered to immigrant students should be bilingual in format. Even if the effect of culturally biased items is small, as was that found by Tsang (1976), it should not be allowed to distort test performance. Many immigrant and refugee students are unfamiliar with the formats and task demands of the various mathematics tests used in American schools. Educators should be sensitive to possible cultural biases when administering tests to such students. Items whose content is alien could be eliminated, or the students could be familiarized with the content before taking the test. Training in test-taking strategies can help students overcome some of the difficulties posed by unfamiliar content and format.

Lastly, the research literature on the mathematics education of Asian-American students and other minority students is sparse. Support for more research is sorely needed to examine the differences in achievement and to design mathematics curricula and instruction appropriate for the educational needs of diverse student groups.

REFERENCES

Bagasao, P. Y. (1983). *Factors related to science-career-planning among Asian and Pacific American college-bound high school seniors.* Unpublished doctoral dissertation, University of California at Los Angeles.

Bernal, E. M., Jr. (1972). Concept learning among Anglo, black and Mexican-American children using facilitation strategies and bilingual techniques. *Dissertation Abstracts International, 32,* 6180A. (University Microfilms No. 72-15707).

Bureau of the Census. (1981). *1980 Census of Population Supplementary Report Race of the Population by States: 1980* (Supplementary Report No. PC80-S1-3). Washington, DC: U.S. Department of Commerce.

Bureau of the Census. (1984). *1980 Census of Population, Vol. I, Characteristics of the population, Chapter D, Detailed population characteristics, Part I, United States Summary, Section A: United States* (Report Number PC80-1-D1-A). Washington, DC: U.S. Department of Commerce.

Ford Foundation. (1983, August). *Refugees and migrants: Problems and program responses.* New York: Author.

Ginther, J. F. (1978). Pretraining Chicano students before administration of a mathematics predictor test. *Journal for Research in Mathematics Education, 9,* 118–125.

Kitano, H. L. (1969). *Japanese Americans: The evolution of a subculture*. Englewood Cliffs, NJ: Prentice-Hall.

Lee, R. H. (1960). *The Chinese in the United States of America*. Hong Kong: Hong Kong University Press.

Ng, K. M., & Tsang, S. L. (1980, April). *Mathematical cognitive structures of Chinese American, Euro-American, and Hong Kong Chinese students*. Paper presented at the meeting of the American Educational Research Association, Boston.

Peng, S. S., Jeffrey, O. A., & Fetters, W. B. (1984, April). *School experiences and performance of Asian American high school students*. Washington, DC: U.S. Department of Education, Office of Educational Research and Improvement.

Ramist, L., & Arbeiter, S. (1984). *Profiles, college-bound seniors, 1982*. New York: College Entrance Examination Board.

Stevenson, H. W. (1983). *Making the grade: School achievement in Japan, Taiwan, and the United States* (Annual Report). Stanford, CA: Stanford University Center for Advanced Study in the Behavioral Sciences.

Tsang, S. L. (1976). The effects of the language factor and the cultural content factor of mathematics achievement tests on Chinese and Chicano students (Doctoral dissertation, Stanford University). *Dissertation Abstracts International, 37*, 714A.

Tsang, S. L. (1983). *Mathematics learning styles of Chinese immigrant students* (Final research report). Oakland, CA: ARC Associates.

Tsang, S. L. (1984). The mathematics education of Asian Americans. *Journal for Research in Mathematics Education, 15*(2), 114–122.

Tsang, S. L., & Wing, L. C. (1985). *Beyond Angel Island: The education of Asian Americans*. New York: Institute for Urban and Minority Education, Teachers College, Columbia University.

U.S. Commission on Civil Rights. (1980). *The tarnished golden door: Civil rights issues in immigration*. Washington, DC: U.S. Government Printing Office.

Chapter 9

Mexican-American Women and Mathematics:
Participation, Aspirations, and Achievement

Patricia MacCorquodale
Department of Sociology
University of Arizona

Concern over the declining technological advantage of the United States has focused attention upon the fundamental math and science training provided by schools. Although basic preparation in math and science for all students needs improvement, the performance of several social groups is especially problematic. Insofar as girls and most minority students are less well prepared in mathematics, their access to higher education, particularly in science, mathematics, engineering, and other technical fields, is severely limited. Society suffers from an inability to utilize the potential of these youth and from increasing economic disparities between members of different racial, ethnic, and gender groups. The crucial issue is identification of the *factors* that discourage female and minority students from obtaining adequate preparation in mathematics.

This chapter examines gender and ethnic differences in math achievement and the explanations offered for those differences. Recent research is reviewed in order to assess the empirical support for the proposed explanatory factors. In addition to information from national and local studies, data from students in southern Arizona are used to determine the contributions of cognitive, cultural, and educational factors to math achievement, participation, and interests.

GENDER AND ETHNIC DIFFERENCES IN MATH ACHIEVEMENT

Recent studies of math achievement have involved gender comparisons or race and ethnic contrasts; however, few inquiries examine gender and ethnicity simultaneously to determine whether minority women are double disadvantaged. The results from a number of studies show that Hispanic, native-American, and black students perform at a lower level than white and Asian-American students. Data from the National Center for Education Statistics in 1978 revealed that Hispanic students scored significantly below majority students in math achievement and that the gap was greater for older compared to younger students (National Center for Education Statistics, 1980).[1] Information from the National Assessment of Educational Progress indicates that the slight decline in the performance of white male students has been offset by a similar decline in the performance of 17-year-old Hispanic students between 1973 and 1978 (National Assessment of Educational Progress, 1979, 1980). The scores of Hispanic students in the younger cohorts remained stable (see Barrow, Mullis, & Phillips, 1982). The consistency and stability of these findings suggest that differences between racial and ethnic groups are unlikely to change in the future unless specific efforts are made to improve the performance of minority students.

Gender differences vary depending upon the specific aspect of math examined and the sample used. Studies utilizing large national samples, such as the National Assessment of Educational Progress or the Scholastic Aptitude Test, find that boys score higher on standardized achievement tests than girls; because a large portion of this difference is due to greater course taking on the part of boys, the achievement gap increases throughout high school. In terms of specific measures, males perform better on problem-solving items, spatial relations, geometric reasoning, and measurement; females score higher on computation problems and simple word problems. In spite of their lower achievement test scores, girls receive higher grades in math classes than males. In the few cases when gender differences are examined in racial and ethnic groups, the results indicate that although race and ethnic differences are greater than gender differences, minority women are affected by both characteristics. (See Fleming & Malone, 1983, and Smith & Lantz, 1981, for information on blacks, Rendón, 1983, and Creswell, n.d., on Hispanics.)

There are a number of definitional problems involved in comparisons of racial and ethnic groups. First, race and ethnicity usually are measured by students'

---

[1]The term *majority student* is used to refer to non-Hispanic, white students, often called Anglos in the Southwest. This group is contrasted to the minority students consisting of several racial and ethnic groups (black, Asian-American, native American, Hispanic). Although non-Hispanic, white students may not be the numerical majority at a particular school, most of the population of our country can be included in this category. The non-Hispanic, white group is used for the purposes of comparison in spite of its diverse representation of ethnic groups (e.g., Polish, French, Irish, etc.) because these distinct groups are similar to each other in their current social position, share a history of early migration to this country, and are too small to analyze separately. Although this practice glosses over differences within the majority group, the categorization of minority students also groups together heterogeneous and distinct populations (e.g., Koreans, Japanese, and Chinese).

self-identification. Not only does the limited number of categories used mask the diversity within any group, but self-identification often is assumed to be equivalent to other dimensions of ethnicity. For example, as De Avila points out in this volume, Hispanics are often assumed to be bilingual, and research findings then are interpreted in terms of language while other dimensions of ethnicity, such as recency of migration, cultural values, and family and community bonds, are not explored. Such simplistic interpretations of ethnic effects are facilitated by use of limited comparison groups; for example, a performance difference between Hispanics and non-Hispanic whites may be attributed to bilingualism but inclusion of a French Canadian group might indicate that the effect is specific to those who speak Spanish. This unidimensional approach to race and ethnicity both limits our understanding of how race and ethnicity affect achievement and reduces our ability to separate the effects of socioeconomic factors from those of culture.

Similarly, gender is often conceptualized simplistically. Because gender is a dichotomous categorization, the inference from gender comparisons is that the sexes are internally homogeneous groups that are distinctly different from each other. In fact, there may be considerable overlap in the distributions on any particular measure, and because of the diversity within each gender, there may be more variation within each gender than between the sexes. For example, contrasts between boys and girls obscure differences between schools where girls' enrollment in advanced math exceeds that of boys in some educational institutions. The oversimplification of gender differences encourages biological explanations and fosters acceptance of the differences and the creation of "remedial" programs. The development of more complex models that include social and educational factors places the learning of math in a broader context and suggests strategies that facilitate achievement across a diverse cross-section of students.

These restricted conceptualizations create special problems for the study of ethnicity and gender. Because most research examines ethnicity or gender, knowledge about how the two factors operate together is severely limited. For example, although we know that girls' interest and enrollment in math is reduced by their view that math will not be useful in their adult lives, we do not have information on the perceived usefulness of math from minority students of both genders who often come from lower socioeconomic backgrounds.[2] Examining both gender and ethnicity raises the issue of whether black girls see math as more useful than black boys because they realize the necessity of their future em-

---

[2]Socioeconomic status or background refers to the position of an individual in the hierarchical organization of society. It is usually measured by income, education, and occupation since most of the stratification in modern industrial societies is based upon type of employment. In the case of children, the socioeconomic characteristics of the parents are used insofar as they influence the standard of living and lifestyle of the entire family. Although popularly referred to as "class," social class has a distinct meaning in social theory.

ployment. Or one might explore whether Hispanic girls are doubly disadvantaged with respect to the perceived utility of math because of their desire to be housewives. Thus, including both gender and ethnicity in research on mathematics compels asking new questions and seeking new explanations of the processes that influence achievement.

A STUDY OF GENDER AND ETHNICITY IN SOUTHERN ARIZONA

This chapter reviews studies of Mexican Americans in order to assess the extent of gender and ethnic differences and their causes. Because only a few studies examine both gender and ethnicity, data from students in the Southwest are presented when applicable. Because of the unique characteristics of minority groups, comparisons are limited to Mexican-American and Anglo (non-Hispanic, white) youth. Although this focus does not provide information about other ethnic groups, specifying the processes through which gender and ethnicity affect achievement in these two groups is a first step that can be replicated in other ethnic and racial groups.

Because this chapter uses data from high school and junior high students in southern Arizona, a brief description of that research is in order.[3] Information about school performance and social support for education was gathered by questionnaires distributed during required classes to high school and junior high students in six schools in southern Arizona. A total of 2,442 questionnaires were completed.[4] The bilingual questionnaire included questions about educational and occupational aspirations, attitudes toward adult roles and school subjects, social support for education, family and background characteristics, and self-image.

The six schools studied were selected on the basis of the proportion of minority students. Students who identified themselves as Mexican or Mexican American were the modal ethnic group in each school. In this chapter, students who identified themselves as Mexican or Mexican American (together equaling 1,589 respondents) are contrasted with those who identified themselves as Anglo ($N = 476$). Students of other ethnicities are excluded from the present discussion.

The research sites vary in terms of urban or rural location and socioeconomic status. The first site utilized the only high school and junior high in a small town of 12,000 that borders a city of nearly 53,000 in Mexico. Eighty-nine percent of the students are Mexican-American. Because there is only one high school and

---

[3]More detailed information about the research design, sampling, and results of this study is available in MacCorquodale (1984).

[4]The response rate was quite high; at one school, less than 1% refused to participate and the highest rate of nonparticipation was 13%.

one junior high for this community, the entire range of socioeconomic background is represented. The other schools were drawn from the largest school district in Arizona, which has an urbanized student population that is 35% minority; the majority of that group is Mexican-American. Within this district there is wide variation in social class. However, because research sites were chosen on the basis of minority enrollment, students from the urban schools overrepresent the lower end of the distribution of parents' educations and occupations.

MATH ACHIEVEMENT: THE EXTENT OF THE DIFFERENCES

Math achievement has been measured using standardized achievement tests, course enrollments, and grades. Data from large national samples indicate that Hispanic and Anglo males score higher than females on standardized math achievement tests (Fernandez & Nielsen, 1986); information from smaller, regional samples find ethnic but no sex differences (Creswell, n.d.; Espinoza, Fernández, & Dornbusch, 1977). On specific measures, males from both groups receive higher scores for arithmetic reasoning than females; Anglo boys demonstrate greater math knowledge than Anglo girls whereas there is no gender difference among Hispanics (Moore & Smith, 1985). Mexican Americans also enroll in fewer math courses in high school than Anglo students (Espinoza et al., 1977; Hansen, 1979; Rendón, 1983). Rendón (1983) found that although Anglo male community college students had taken more high school math courses than Anglo females, among Hispanics there was no sex difference.

Data from southern Arizona show the cumulative effect of course-taking patterns. As shown in Figs. 9.1 and 9.2, the effects of sex and ethnicity are not linear, but vary with grade level. In other words, the differences between the sexes and ethnic groups are small initially but increase over time. These increasing differences suggest that the causes are not due to stable factors, such as biology, but rather are based upon factors that vary throughout school, such as encouragement or motivation. The changes over time are also related to the structured nature of the math curriculum.[5] Math is required of seventh- and eighth-grade students in both districts, and therefore, the sex and ethnic differences are quite small in junior high. Only 1 year of math is required for high school graduation although 2 years are recommended for students planning to attend college. Once the element of choice is introduced, sex and ethnic differences increase. Although most students take two or three math courses in high

---

[5]Grade in school is the strongest predictor of number of math courses taken accounting for 30% of the variation.

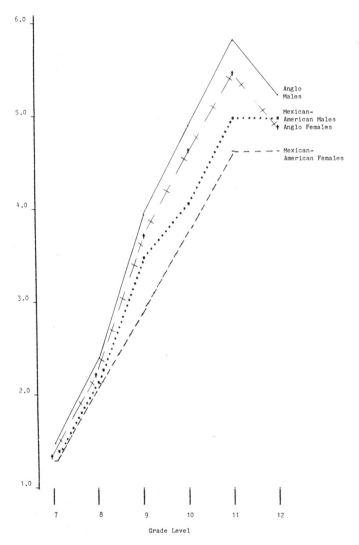

FIG. 9.1. Number of math classes taken by grade, sex and ethnicity: rural, border district.

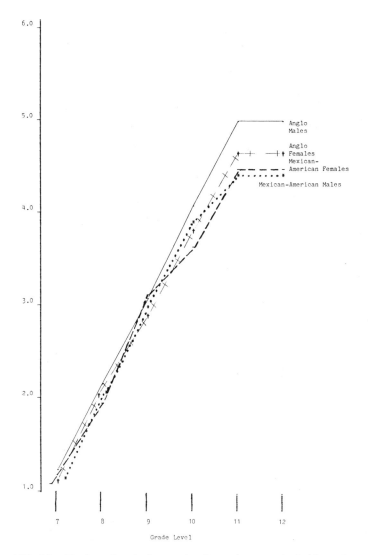

FIG. 9.2. Number of math classes taken by grade, sex and ethnicity: urban district.

school, less than 6% take math all 4 years. By the 11th grade Anglos have completed more math than Mexican Americans and boys more than girls.

Enrollment patterns in specific courses pinpoint the effects of sex and ethnicity. As shown in Table 9.1, the differences are minimal in advanced electives, which are taken by a small percentage of the student body. Patterns of participation in basic courses vary significantly by sex and ethnicity. Anglo students are more likely to take Algebra 1, and Mexican Americans are more likely to enroll in general math. Since Algebra 1 is preferred course for fulfilling graduation requirements, these ethnic differences suggest that Mexican Americans take a less specialized course to meet requirements or need to fulfill prerequisites before they can take algebra. In either case, the opportunities for Mexican Americans to continue taking math, to take science courses that require algebra, and to have adequate preparation for higher education are significantly reduced.

Geometry, another basic course, is more likely to be taken by boys and Anglos. The link between geometry and spatial visualization skills merits further attention; of particular interest is whether the less developed spatial visualization skills shown by girls and Mexican Americans lead them to avoid geometry, whether their decisions not to enroll in geometry reduce their spatialization skills, or some combination of both. The importance of language skills in learning geometry also merits investigation insofar as language skills may be more necessary for understanding geometry than other math courses that rely on computation and abstract symbols. Boys are more likely than girls to take pre-

TABLE 9.1
Course Enrollments by Gender and Ethnicity

Percent who have taken:	Anglo Males	Anglo Females	Gender Diff.*	Mex.-Am. Males	Mex.-Am. Females	Gender Diff.*
7th-grade math	94.4	92.4	+2.0	90.0	89.5	+.5
8th-grade math	64.5	67.0	−2.5	73.1	71.5	+1.6
Pre-algebra	32.5	28.8	+3.7**	31.5	25.2	+6.3
Algebra I	44.0	59.3***	−15.3	45.9	42.1	+3.8
Algebra II	16.2	13.1	+3.1	17.5	15.0	+2.5
Geometry	30.3	22.5***	+7.8**	19.8	17.0	+2.8
College algebra	5.1	1.7	+3.4	1.9	2.2	−.3
Calculus	1.3	.4	+.9	.9	.2	+.7
General math	32.9	27.1***	+5.8	42.8	39.8	+3.0
Business math	7.7	3.8	+3.9	7.5	7.1	+.4
N =	239	237		755	834	

Note: *   gender difference = (male − female)
      **   significant gender difference $p < .01$
      ***significant ethnic difference $p < .01$

algebra, a junior high course offered as advanced preparation for high school algebra. Boys may be more likely to anticipate their future course work, or they may be encouraged by parents, teachers, and counselors to consider their high school plans earlier than girls, a factor discussed here.

The table also illustrates the importance of examining sex and ethnicity simultaneously because the effects of gender are not uniform across the subgroups. For example, among Anglos, girls are more likely than boys to have completed eighth-grade math and Algebra 1. The gender difference for pre-algebra is much larger among Mexican Americans than Anglos, but gender differences are larger for Anglos with respect to geometry, general math, and business math. Thus, the failure to differentiate students on the basis of ethnicity reduces the size of the gender effects.

The issue of grades is more complicated; Mexican Americans receive lower grades than Anglos in high school mathematics courses, but girls of both ethnicities report higher grades than boys (Espinoza et al., 1977; Rendón, 1983). For example, in southern Arizona, 27% of the girls but only 21% of the boys report getting mostly A's in math (MacCorquodale, 1984). In an analysis that controlled for socioeconomic status and race, course work and grades affected math achievement such that if girls did not outperform boys in math courses, the sex difference on achievement tests would be even greater (Pallas & Alexander, 1983). Other research has shown that ability is more strongly related to achievement among boys than girls (Ethington & Wolfe, 1986) and for Anglos compared to Hispanics (Crisco, 1975; Evans & Anderson, 1973). Girls, especially the most highly able, underrate their ability when choosing a major, and therefore, are less likely to enter math, science, and engineering (Parelius, 1981). These findings suggest that social influences discourage girls and minority students from translating their abilities into achievement in mathematics throughout their educational careers.

CURRENT EXPLANATIONS OF GENDER AND ETHNIC DIFFERENCES

Cognitive Style

Interestingly, many of the factors used to explain gender differences in mathematics are also offered to account for racial and ethnic differences. For example, both women and minorities use cognitive styles that are more field dependent and score lower on measures of spatial visualization than boys and majority students (Fennema & Sherman, 1977; Ramirez & Price-Williams, 1974). Spatialization skills are related to math achievement and to general measures of intelligence (Roberge & Flexer, 1981; Satterly, 1976) and over time these skills are more important for the math achievement of girls than of boys (Fennema & Sherman,

1977; Sherman, 1980). Although spatialization skills may aid in learning math, particularly geometry, they are not a sufficient explanation of ethnic and gender differences insofar as there are other ways to solve problems and other influences on math achievement and problem-solving ability (see Fennema, 1981). More important, research has not demonstrated that spatialization is associated with math achievement across ethnic groups, that gender effects are the same across ethnic groups, nor that controlling for spatialization skills reduces sex and gender differences in mathematics course taking.

Discussions of cognitive style lead to debates about the biological bases of differences (Wittig & Peterson, 1979). In particular, sex differences in brain structure have been suggested as a factor influencing cognitive style. Logically, it is hard to imagine biological factors common to Hispanics and girls that would account for differences of cognitive style. Evidence that cognitive styles change over time (Sherman, 1980) can be altered with intervention programs (Fox, Brody, & Harrop, 1979; Keene, 1978; Kreinberg, 1978), and vary with acculturation among Hispanics (Ramirez, Castañeda, & Herold, 1974), support a social explanation of the origins of these gender and ethnic differences.

Motivation

The perceived usefulness of mathematics is extremely influential in decisions to take more mathematics for both girls and minority students (Armstrong, 1980; Espinoza et al., 1975). Additionally, the more important a subject is rated in terms of obtaining and succeeding at work, the greater the effort students report expending (Espinoza et al., 1975). The degree to which math is seen as useful is dependent upon the career and educational aspirations of students. Although minority and female students place high value on education and desire a college education in proportions similar to majority and male students, their interest in specific careers that use math influences their academic achievements and motivation. Different factors influence the career interests of female and minority students (Berryman, 1983). Although girls do not differ from boys in mathematical achievement upon entering high school, girls like mathematics less and are less interested in math-related careers than boys (Wise et al., 1979). Hispanic students' interest and aspirations are reduced by the low level of parental education and early pre-college tracking (Berryman, 1983). That Mexican-American women are affected by both sets of factors is shown by their declining share of post-high-school degrees in mathematics from 1975 to 1979 and their stable level of college attainment for the last 20 years (Berryman, 1983).

Data from students in southern Arizona indicate that differences in achievement cannot be explained by a lack of interest in math on the part of minority or female students. On a variety of measures, including favorite subject, interest in taking math, caring about math grades, trying to do better in math, and the importance of math for future work, understanding the world and consumer decisions, girls and Mexican Americans have more positive attitudes than boys and Anglos (MacCorquodale, 1983). Multivariate analyses indicate that although many of these variables are related to the number of math courses taken, the positive attitudes of Mexican-American and female students should lead them to take more courses than their Anglo and male counterparts.[6]

Students' motivation is influenced by the meanings given to academic performance. Attribution research has found gender and ethnic differences in the reasons students give to explain their behavior; Hispanics are more likely than Anglos to believe that their successes and failures are based upon ability and effort (Powers & Wagner, 1983). This reasoning leads students to conclude that stable factors, internal to the individual, are responsible for academic performance (as compared to luck, teachers' attitudes, difficulty of tasks, etc.). These internal explanations have contradictory effects upon motivation. When the outcome is success, self-esteem is raised because effort and ability pay off, but when the result is failure, students blame their own lack of ability or effort. Similarly, when choosing careers, girls are more likely to mention enjoyment and interest, whereas boys give goal-directed, practical reasons; Hispanics base their decisions on perceived ability more than Anglos or blacks do (Block, Denker, & Tittle, 1981). Thus, the meanings of performance and ability should be investigated as influences on math achievement and participation.

Students' views of themselves are indirect measures of motivation insofar as self-image influences, and is influenced by, feedback received from others in daily life. Since students' time is spent primarily in school, characteristics that are related to academic performance are likely to influence motivation and achievement. Seeing the self as intelligent is particularly important; it is related to interest in taking science and math and to number of math courses taken.[7] Among Mexican Americans, the association between interest in math and intelligence is curvilinear (MacCorquodale, in press). This finding suggests that among Hispanics, feedback is disregarded when determining interests, but that

---

[6]Regression analyses were performed with different sets of related variables in order to determine which had independent effects in explaining the number of math courses taken. Because the details of the analyses may not be of interest to the general reader, the results are summarized in this chapter; equations and other specific information are available from the author. The results indicate that (a) caring about math grades, math grades, trying to do well in math, (b) importance of math for understanding the world, (c) whether science is the favorite class and whether math is least favorite class are significantly related to number of courses taken. Each set explains about 1% of the variation in course taking.

[7]Interest in taking math courses is related to ratings of creativity in a curvilinear manner; those students who see themselves moderately creative are more interested in math than those who are high or low on creativity. This interest is not translated into course taking, which is positively related to math enrollments. Multivariate analyses indicate that the number of math courses taken in high school is related to ethnicity and ratings of intelligence, independence, creativity, and competitiveness. Together these variables explain about 7% of the variation in number of courses taken with intelligence and ethnicity contributing most to the regression equation.

course taking is affected by perceptions of the self as intelligent. That parents, teachers, and counselors are important sources of information, particularly with respect to intelligence, demonstrates the role of social factors in the development of self-image (Eccles & Jacobs, 1986; MacCorquodale, 1986). Thus, girls' and Mexican Americans' views of themselves as less intelligent than boys and Anglos reflect the feedback they receive from their social environment and influences their aspirations and achievements.

Research in Arizona on sex role attitudes reveals that sex and ethnic differences are not consistent. Girls, for example, are more likely than boys to believe that "boys do not like girls who do better than they do in math or science." Boys are more likely than girls to agree that "most men would not want to marry a woman who was interested in becoming a scientist or mathematician." Boys and Mexican Americans believe more strongly than girls and Anglos that "men need to know more math than women." Girls' sex role attitudes and their perception of boys' attitudes influence course-taking patterns and indicate one of the ways that Mexican-American girls are doubly disadvantaged in the study of mathematics.

Educational and career aspirations influence course taking among high school students in southern Arizona. Both ethnicity and gender influence students' interest in science and math careers. Compared to 4.2% of the Anglo and 4.8% of the Mexican-American girls, 7.9% of the Anglo and 5.8% of the Mexican-American boys are interested in math-related careers. Together ethnicity, interest in a math-related career, and the importance of math to future work and understanding the world are among the strongest predictors of course taking in high school.[8] Evidence from studies of science indicate that boys are more likely than girls to explore their academic interests outside of the classroom (Kahle, Matyas, & Cho, 1985). Thus, motivational interests may be reinforced by extracurricular experiences. Because career interests and sex role attitudes are formed in elementary and middle school, efforts to alter the aspirations of minority and female students must be designed for these educational levels. Longitudinal studies are needed in order to separate those factors, both within and outside of the classroom, that influence the setting of educational and occupational goals from those that determine whether students are able to meet their goals.

Language

Although the assumption that Mexican Americans have a "language problem" that interferes with education has been repeatedly challenged (Carter & Segura, 1979), the role of language in academic achievement continues to be debated.

---

[8]In a regression analysis, these variables explain 7% of the variance in the number of math courses taken by high school students.

There are two issues related to language. First, does inadequate preparation in English impede the learning of mathematics? Because most classroom instruction relies upon English, those students who speak only limited English have lower achievement in all academic areas including math (Espinoza et al., 1975; Sells, 1980). Those students with deficiencies in English or who are studying English as a second language have course conflicts because of the necessity of learning English before they can take more math and science. The structure of the curriculum also limits options for studying math, particularly for those who must take more English, because more credits in English than math are required for high school graduation.

The second issue focuses upon bilingualism. The debate centers around whether students who are bilingual develop cognitive systems that facilitate the learning of mathematics. Although it is argued that math is the subject most independent of language and culture (Serna, 1980), others propose that bilingualism facilitates the development of more heterogeneous, diversified mental abilities, superior patterns of concept formation, and greater flexibility in task orientation (Diebold, 1982; Ramirez, 1977). The empirical evidence is also mixed; research shows both that bilinguals make more errors in verbal math problems (Mestre, Gerace, & Lochhead, 1982) and that they score no differently than students who speak only English (Fernandez & Nielsen, 1986; Sells, personal communication, 1980). Use of Spanish in the home is negatively related to achievement (Fernandez & Nielsen, 1986) but positively related to grades (Dornbusch, Fraleigh, & Ritter, 1986). Further research is necessary in order to specify the relationship between bilingualism and cognitive skills. In terms of the teaching of math, the paucity of bilingual material for elementary school mathematics (Gallegos, 1980) and the limited offerings in middle and high school (MacCorquodale, 1984) reduce the possibilities of learning math in a bilingual setting.

One of the problems in the study of language is the confusion of language with other processes that also affect learning. The effects of language, for example, should be differentiated from other characteristics of the culture (see Saxe's chapter) and from patterns of interaction within classrooms (see De Avila's chapter). Cultural characteristics, such as use of Spanish in the home or celebrations of traditional holidays, may be confounded with socioeconomic status. Because socioeconomic status is related to college attendance, math and verbal achievement, and career interests (Berryman, 1983; Fernandez & Nielsen, 1986; Mestre & Robinson, 1983), it is necessary to control for socioeconomic status in analyses of language and cultural effects. Conceptual clarity is especially important when studying gender and ethnicity because although the effects of language should not affect boys and girls differently, the influences of culture and classroom interaction may be gender specific.

Multivariate analyses of data from southern Arizona youth were used in order to distinguish between the effects of language, culture, and socioeconomic status. Stepwise regression analysis was used in order to determine which variables were associated with proclivity toward taking math courses, independent of the relationships between the variables. For example, since speaking Spanish is more common in families of lower socioeconomic status, this analysis would indicate whether language affects math enrollments after controlling for socioeconomic status.

The results presented in Table 9.2 indicate that four factors have independent effects on taking courses in math. First, proficiency in English and not bilingualism (as measured by language spoken at home or with friends) is the strongest predictor of number of courses taken. Second, contrary to what might be expected, after controlling for the other factors, mothers born in Mexico or Latin America have children who take more math. Fernandez and Nielsen (1986) also found that recent immigrants did better on achievement tests than those with longer residence. This pattern is consistent with Dornbusch, Fraleigh and Ritter's conclusion that those who speak a language other than English at home, assumed to be more recent immigrants, have higher achievement motivation than those who are more "assimilated." Socioeconomic status has an additional effect; students whose fathers and mothers have more education take more math. Other research has found that parental education is positively related to high school grades (Dornbusch et al., 1986) and that minority students with college-educated parents are more likely to attend college and to pursue science and math interests (Berryman, 1983). Finally, the inclusion of gender shows that English grades,

TABLE 9.2
Math Course Taking, Language, Culture
and Socioeconomic Status:
Stepwise Multiple Regression Analysis

Number of Math Courses Taken in High School
(Mexican-American students only)

Variable*	Standardized Beta Weight	Change in R^2
Grades in English	.314	.105
Father's education	.086	.013
Sex	−.082	.007
Mother foreign born	.093	.005
Mother's education	.088	.006

Note: Multiple R = .369
$F = 29.69 (5,942) p < .0001$
$N = 1,077$
   *variables not significant in the equation: language spoken at home, language spoken with friends, student foreign born, father foreign born.

socioeconomic status and generation do not account for the differences between the sexes.

In summary, these results indicate that the language effects on math are due to skill in English rather than language usage outside of school. After controlling for grades in English, socioeconomić status and generational status additionally influence math course taking; those students whose parents have less education and longer residence in the United States take less math. The issue that remains to be explored is the processes through which these characteristics of parents, which represent their position in the social structure, affect educational outcomes of children.

Familism

Among a variety of cultural characteristics, familism has received the most attention in explaining achievement. Familism refers to the relative importance of family members in determining an individual's values, goals, and orientation. Although there is consensus that the family is important in Mexican-American culture, the consequences of this family orientation on education continue to be debated. One viewpoint describes the family as hindering intellectual development because the needs of the family as a whole are seen as superseding the needs of any particular family member (Grebler, Moore, & Guzman, 1970; Montez, 1960). For this reason, the principal dilemma facing Mexican Americans is framed in terms of educational and occupational achievement versus fulfillment of traditional family roles (Dworkin, 1968). For example, Mexican Americans with greater independence from their families had greater educational achievements than those who maintained their dependence (Schwartz, 1971).

The contrasting view portrays the family and achievement as congruous. The warm, collectivist atmosphere of the family is seen as providing the support necessary to strive toward goals and adapt to changes in other social institutions (Gonzales, 1979; Mirandé, 1977; Zinn, 1980). Empirical studies support the view that family orientation does not interfere with educational aspirations and achievement (Lopez-Lee, 1972); thus Mexican Americans and Anglos do not differ in their educational values and aspirations (Aiken, 1979; Espinoza et al., 1977; Juarez & Kuvlesky, 1969). Hispanic students who identified more with the Mexican-American culture were more educationally minded and more academically successful than those with less identification (Lopez-Lee, 1972; Ramirez, 1971). However, family orientation might help explain gender differences insofar as girls may be given more family responsibilities while still in school and may limit their aspirations in order to fulfill family obligations.

Family responsibilities of students do not appear to be related to educational achievements or aspirations. Mexican Americans are less likely than Anglos to be involved with household tasks, primarily because their mothers take more exclusive responsibility for housework. They are more likely than Anglos to do childcare, primarily because they come from larger families. And although they do more childcare, Mexican Americans are more likely than Anglos to do homework and Mexican-American girls spend more time doing homework than their Anglo counterparts (MacCorquodale, 1984). Family obligations, however, appear to influence higher education and career options since Mexican Americans are more likely to believe that they will have to stop school before they complete their desired education due to family duties (MacCorquodale, 1984). Anglos, on the other hand, are more likely to believe that they will be able to meet their educational aspirations. These perceptions are related to math enrollments because those who are more likely to believe they will complete their education take more math. Mexican-American girls, in particular, may limit their education beyond high school since they expect to have larger families and are more likely to believe that women's work and family roles conflict compared to Anglo girls.

The behavior and values of family members also influence educational goals. Parents are particularly important influences on the aspirations of Mexican-American youth (Schwartz, 1971; TenHouten, Tquejen, Kendall, & Gordon, 1971; Webster & Fretz, 1978). In particular, parental encouragement to enter a math or science field is a key factor among community college students (Rendón, 1983). In southern Arizona, parents from both ethnic groups value education highly and encourage their children to get a college education. In fact, Mexican-American students believe that a college education is more important to their parents than Anglo students do. The importance to parents of college is related to taking math courses in high school, but does not eliminate the ethnic difference.[9] The social context also provides encouragement and support for the learning of specific subjects for all students. Mexican Americans, however, report that learning math is slightly more important to their mothers, fathers, and friends and that they get more encouragement to do well in math than Anglo students. Multivariate analyses indicate that the importance of math to parents is related to course taking in high school and encouragement from mother is related to course taking in middle school and high school. Generally, these effects are small and do not compare with the importance of college education to parents.

Given that Mexican-American parents are supportive and encouraging of their children's education, why then do ethnic differences in educational attainment and math achievement persist? The answer lies in parents' ability to transform their encouragement into supportive, concrete actions. For example, Anglos are more likely to get help with their math homework from fathers, and Anglo boys get more help from their mothers than Mexican-American boys (MacCorquodale, 1983). In part because of their own limited educational background,

---

[9]Taken together, reasons for stopping school, importance of college to parents, and ethnicity account for about 8% of the variance in math course taking for high school students. Of these, the importance of college is most significant, accounting for 5% of the variance.

Mexican-American parents have different views of the educational system than Anglo parents.[10] Mexican-American parents are more likely than Anglos to believe that schools do a good job of preparing students for education beyond high school.[11] They also report more attendance at parent-teacher conferences and extracurricular school events and more frequent conversations with students about school performance and daily activities. Evidence from California indicates that parental participation in school programs has a positive effect on grades (Dornbusch et al., 1986). When parents in Arizona were asked what advice they would give from their own educational experience, Mexican Americans were twice as likely as Anglos to stress the importance of education for success and three times as likely to suggest working hard. Anglo parents were more likely to give strategies for achievement (stay in school, plan carefully, get a respectable profession) or affective advice (relax, do what you like).

In terms of specific advice or action, Mexican-American parents are less able to apply their diffuse support. Among Anglos 56% of the parents of boys and 67% of girls' parents help their children select courses. For Mexican Americans, the comparable figures are 29% and 44%. Mexican-American parents believe that they do not have enough information to help in course selection, and when asked how much math is required for high school graduation, three-quarters would not hazard a guess. Only a third of the Anglo parents did not offer an answer. When asked to recommend occupations that their son or daughter should consider if they were interested in math, nearly a quarter, compared to about 10% of the Anglos, did not know any possible occupations. Anglo parents gave a broader range of occupations and more high-status jobs. Mexican-American parents' answers were concentrated in accounting and data processing whereas engineering and business were more popular among Anglo parents.

Another issue under the topic of familism, parental sex role attitudes, partially explain sex and ethnic differences in math course taking and career interest (Eccles & Jacobs, 1986). Mexican-American parents are, generally, more traditional in their attitudes toward sex roles. Although the vast majority of all parents hold liberal attitudes as to women's participation in math, science, and engineering, attitudes about role conflict for employed women are more traditional. For example, 25% of the Anglo parents and 35% of the Mexican-American parents believe that women with elementary-school-aged children should stay home rather than work; 37% of the Anglos and 44% of the Mexican Americans agree that working wives are too independent. Thus, Mexican-American girls are

---

[10]Information from parents comes from interviews with 258 parents of randomly selected 8th-, 10th- and 12th-grade students. More information about the parent sample and interviews is available in MacCorquodale (1984).

[11]This finding conflicts with research comparing recent immigrants with Chicano families which found that Chicano parents were the most alienated from the school system and believed that their children were not getting an adequate education (Romo, 1985).

likely to be discouraged from pursuing nontraditional careers insofar as their parents are more likely than Anglo parents to see role conflicts for women who attempt to combine work and family life.

In summary, Mexican-American parents are quite supportive of education and math, have higher expectations for their children's achievements than Anglo parents, but lack the experience and information necessary to assist their children's educational efforts. One strategy for increasing Mexican Americans' interest in math is to provide parents with more information about educational requirements, course choices, and career options. Information on the percentage of women in the paid labor force and the potential remuneration for women in nontraditional careers, especially science, engineering, and math-related jobs, would give parents an informed basis for advising their daughters. The need for such information is highlighted by the effect on parents of media reports of genetic explanations for gender differences in math aptitude (Eccles & Jacobs, 1986). More research is needed to identify the kinds of information that are valuable to parents, which information and programs are most effective in helping parents channel their support in specific, direct action, strategies to get the information to parents, and parents' receptiveness to the expansion of the schools' influence into an area of family responsibility.

Schools

Classroom dynamics and attitudes of school personnel have received less attention than student characteristics and behaviors despite evidence that teachers' control and structure of the classroom are related to morale, learning, motivation, and achievement (De Avila, this volume; Pascarella, Walberg, Junker, & Haertel, 1981). The attitudes of teachers and counselors are particularly important influences on the interests, motivations, and aspirations of minority students because minority parents may be limited in the kinds of help they can provide. Because teachers work within the larger social context, they often reinforce dynamics that have created differences between the sex and ethnic groups. First of all, teachers accept societal stereotypes about sex differences and ethnic characteristics (Brischetto & Arciniega, 1973; Hernández, 1973) and these stereotypes shape teachers' behaviors. Classroom observations reveal that teachers react more to racial and ethnic characteristics than to gender and alter their expectations of student's scholastic performance accordingly (Wiley & Eskilson, 1978). Although Mexican Americans who use standard English are treated more like Anglo students (Laosa, 1977), significant differences between monolingual speakers remain that are due to reactions to students on the basis of ethnicity and social class (Fernandez & Nielsen, 1986). Because of low expectations for minority students, teachers often ignore Mexican-American children (Jackson & Cosca, 1974), are friendly and warm but do not provide concrete assistance to Chicano and black students (Espinoza et al., 1977). High school enrollments in math for all students are related to teachers' help with math homework in southern Arizona. The importance of interactions with teachers for minority students is underscored by Hispanic community college students who desire personal contact with math and science teachers, outside class contact, and tutorial assistance more than Anglos do (Rendón, 1983).

Second, teachers often share sex-stereotyped attitudes about female participation in nontraditional careers. Teachers reinforce boys positively and girls negatively in technical courses; such reinforcement apparently derives from the well-accepted belief that boys are better in math and science than girls (Burton, 1979; Ernest, 1976; Fox, 1977). Information from teachers in southern Arizona indicates that teachers' attitudes, like those of parents, support math, engineering, and science for women while remaining more traditional with respect to the problems associated with women combining marriage and paid employment. These findings suggest that recent efforts to change attitudes about math as a male domain have been successful; now attention needs to be directed toward attitudes about careers for women and role conflict because traditional attitudes on the part of students, teachers, and parents discourage girls from translating their interests and abilities into occupational success.

Counselors play a key role for minority students in providing information about courses and careers that students and parents lack (MacCorquodale, 1984; Pallone, Hurley, & Rickard, 1973). Counselors' influence is affected by their attitudes and the time available to interact with students. Counselors, for example, evaluate the role of career women in a pattern that is unrelated to the role of mother and average woman; the corresponding roles for men are not seen as conflicting (Ahrons, 1976). Counselors regularly steer girls away from courses in mathematics and science with the well-intentioned belief that such courses are inherently more difficult for girls than literature, art, and the social sciences (Smith & Stroup; 1978). Data from southern Arizona indicate that counselors have a much stronger effect on math enrollments than teachers. Of primary importance is the frequency with which students discuss future work plans with counselors.[12] The demands upon counselors' time is evident in that half of the students in this research have never talked with their counselor about their work interests although minority students get more such counseling than majority students.

Similarly, those students who get more encouragement from counselors to do well in math take more math courses in middle and high school. Mexican-American students report more encouragement than Anglo students. At the com-

---

[12]Encouragement from counselors to learn math and frequency of vocational planning explain 9% of the variation in math taken by both middle and high school students and 6% of the variation for high school students alone. In addition, teacher's help with math homework, school, and ethnicity have significant effects on math taken by high school students. A total of 11% of the variance in high school math is accounted for by these five variables.

munity college level, students of all ethnicities desire more contact with counselors, more information about math and science careers, assistance in developing study skills and in overcoming math fear, and encouragement for women to
enter nontraditional careers (Rendón, 1983). These results suggest that the gender and ethnic gap in math achievement would be even greater if Mexican-
American and female students were not receiving advice and encouragement
from counselors. Interest in math courses and careers for minority students,
especially minority girls, could be increased by ensuring that all students speak
with counselors about career and educational planning and by increasing the
proportion of time that counselors spend providing such assistance to students.

CONCLUSIONS

The gap between Mexican-American and Anglo students in math achievement is
primarily due to the influence of social forces, including encouragement and
information from parents, teachers, and counselors. The cultural component of
the ethnic difference appears to be proficiency in English rather than bilingualism
or other cultural characteristics. The gender difference is related to sex role
stereotyping, the paucity of role models, and perceptions of role conflict for
employed women. Several sets of explanations do not account for the gender and
ethnic difference, such as student motivation and interest in math, grades, family
orientation, cognitive styles, and parental aspirations. The effects of these variables suggest that there should not be gender and ethnic differences and that
Mexican-American and female students should take more math than Anglos and
boys. Comparison of the explanatory power of these sets of explanations indicates that those factors that discourage girls and Mexican Americans are stronger
than those that encourage them. Future research needs to identify other factors
and to determine how they interact with gender, ethnicity, or the combination of
gender and ethnicity. Detailed examination of the content of interactions with
teachers, counselors, parents, and friends will be essential in specifying how
these social processes operate in daily life.

ACKNOWLEDGMENTS

Data used in this chapter were gathered under a grant from the National Institute for
Education (NIE-G-79-0111). The ideas and opinions presented in this chapter are those of
the author and do not represent the positions or policies of the National Institute of
Education. This chapter was written while the author was a Visiting Scholar at the Center
for Research on Women, Stanford University. Their support and those of my colleagues at
the University of Arizona are gratefully acknowledged.

REFERENCES

Ahrons, C. (1976). Counselor's perceptions of career images of women. *Journal of Vocational Behavior, 8,* 197–207.

Aiken, L. R. (1979). Educational values of Anglo-American and Mexican-American college students. *The Journal of Psychology, 102,* 317–321.

Armstrong, J. (1980). *Achievement and participation of women in mathematics: An overview* (Final report summary). Washington, DC: National Institute of Education.

Barrow, K., Mullis, I., & Phillips, D. (1982). *Achievement and the three R's: A synopsis of National Assessment findings in reading, writing and mathematics.* Denver, CO: American Educational Research Association. (ERIC Document Reproduction Service No. ED 223 658)

Berryman, S. E. (1983). *Who will do science?* (A special report). New York: Rockefeller Foundation.

Block, J., Denker, E. R., & Tittle, C. K. (1981). Perceived influences on career choices of eleventh graders: Sex, SES and ethnic group comparisons. *Sex Roles, 7,* 895–904.

Brischetto, R., & Arciniega, T. (1973). Examining the examiners: A look at educator's perspectives on the Chicano student. In R. O. de la Garza, Z. A. Kruszewski, & T. A. Arciniega (Eds.), *Chicanos and native Americans* (pp. 23–42). Englewood Cliffs, NJ: Prentice-Hall.

Burton, G. M. (1979). Regardless of sex. *Mathematics Teacher, 72,* 293–311.

Carter, T. P., & Segura, R. D. (1979). *Mexican Americans in schools: A decade of change.* New York: College Entrance Examination Board.

Creswell, J. L. (n.d.). *Sex related differences in mathematics achievement of Black and Chicano adolescents.* Unpublished paper.

Crisco, J. J. (1975). *The prediction of Academic performance for minority engineering students from selected achievement-proficiency, personality, cognitive style and demographic variables.* Unpublished doctoral dissertation, Marquette University.

Diebold, A. R. (1982). Consequences of early bilingualism in cognitive development and personality formation. In P. R. Turner (Ed.), *Bilingualism in the southwest* (pp. 29–58). Tucson, AZ: University of Arizona Press.

Dornbusch, S. M., Fraleigh, M. J., & Ritter, P. L. (1986). *A report to the national advisory board of the study of Stanford and the schools on the main findings of our collaborative study of families and schools.* Stanford: Center for Youth Studies.

Dworkin, A. G. (1973). A city founded, a people lost. In I. L. Duran & H. R. Bernard (Eds.), *Introduction to Chicano studies* (pp. 406–420). New York: Macmillan. (Originally published 1968)

Eccles, J. S., & Jacobs, J. E. (1986). Social forces shape math attitudes and performance. *Signs, 11,* 367–380.

Ernest, J. (1976). *Mathematics and sex.* Santa Barbara, CA: University of California-Santa Barbara.

Espinoza, R. W., Fernandez, C., & Dornbusch, S. M. (1975). Factors affecting Chicano effort and achievement in high school. *Astibos: Journal of Chicano Research, 1,* 9–30.

Espinoza, R. W., Fernandez, C., & Dornbusch, S. M. (1977). Chicano perceptions of high school and Chicano performance. *Atzlan, 8,* 133–155.

Ethington, C. A., & Wolfe, L. M. (1986). A structural model of mathematics achievement for men and women. *American Educational Research Journal, 23,* 65–75.

Evans, F., & Anderson, J. (1973). The psychocultural origin of achievement and achievement motivation: The Mexican American family. *Sociology of Education, 46,* 396–416.

Fennema, E. H. (1981). The sex factor. In E. H. Fennema (Ed.), *Mathematics education research: Implications for the 80's* (pp. 92–105). Alexandria, VA: Association for Supervision and Curriculum Development.

Fennema, E. H., & Sherman, J. (1977). Sex-related differences in mathematics achievement, spatial visualization, and affective factors. *American Educational Research Journal, 14,* 51–71.

Fernandez, R. M., & Nielsen, F. (1986). Bilingualism and Hispanic scholastic achievement: Some baseline results. *Social Science Research, 15,* 43–70.

Fleming, M. L., & Malone, M. R. (1983). The relationship of student characteristics and student performance in science as viewed by meta-analysis research. *Journal of Research in Science Teaching, 20,* 481–495.

Fox, L. H. (1977). The effects of sex role socialization on mathematics participation and achievement. In *Women and mathematics: Research perspectives for change* (Papers in Education and Work, no. 8). Washington, DC: National Institute of Education.

Fox, L. H., Brody, L., & Harrop, I. M. (1979). *Women and mathematics: The impact of early intervention programs upon course-taking and attitudes in high school* (Final report). Washington, DC: National Institute of Education.

Gallego, T. A. (1980). Language and culture in the mathematics curriculum of the 1980's: Issues affecting Spanish bilingual students. In Salient issues in mathematics education research for minorities. *Proceedings from a National Institute of Education sponsored meeting* (pp. 16–23). Washington, DC: National Institute of Education (ERIC Document Reproduction Service No. ED 191 699)

Gonzales, S. A. (1979). The Chicana perspective: A design for self-awareness. In D. A. Trejo (Ed.), *The Chicanos: As we see ourselves* (pp. 81–100). Tucson, AZ: University of Arizona Press.

Grebler, L., Moore, J. W., & Guzman, R. C. (1973). The family: Variations in time and space. In I. L. Duran & H. R. Bernard (Eds.), *Introduction to Chicano studies* (pp. 309–331). New York: Macmillan. (Originally published 1970)

Hansen, J. (1979). *A brief survey of Mexican American school achievement in Colorado during the 1970's.* Washington, DC: National Institute of Education. (ERIC Document Reproduction Service No. ED 182 104)

Hernandez, N. (1973). Variables affecting achievement of middle school Mexican-American students. *Review of Educational Research, 43,* 1–39.

Jackson, G., & Cosca, C. (1974). The inequality of educational opportunity in the Southwest: An observational study of ethnically mixed classrooms. *American Educational Research Journal, 3,* 219–229.

Juarez, R. Z., & Kuvlesky, W. P. (1969). *Ethnic group identity and orientations toward educational attainment: A comparison of Mexican-American and Anglo boys.* Washington, DC: National Institute of Education (ERIC Document Reproduction Service No. ED 023 497)

Kahle, J. B., Matyas, M. L., & Cho, H. H. (1985). An assessment of the impact of science experiences on the career choices of male and female biology students. *Journal of Research in Science Teaching, 22,* 385–394.

Keene, F. (1978). *Fundamental algebra and some characteristics of high risk students.* Paper presented at the Conference on the Problem of Math Anxiety, California State University, Fresno, CA.

Kreinberg, N. (1978). *Reducing math avoidance through early intervention and teacher training.* Paper presented at the Conference on the Problem of Math Anxiety, California State University, Fresno, CA.

Laosa, L. M. (1977). Inequality in the classroom: Observational Research on teacher-student interactions. *Aztlan, 8,* 51–67.

Lopez-Lee, D. (1972). The academic performance and attitudes among Chicanos and Anglos in college. *The Journal of Mexican-American Studies, 1,* 201–222.

MacCorquodale, P. L. (1983). *Mexican Americans and mathematics: Aptitudes, attitudes and aspirations.* Unpublished paper.

MacCorquodale, P. L. (1984). *Social influences on the participation of Mexican-American women in science* (Final report). Washington, DC: National Institute of Education. (ERIC Document Reproduction Service No. ED 234 991)

MacCorquodale, P. L. (1986). *Gender and ethnic difference in self-image*. Reno, NV: Western Social Science Association.

MacCorquodale, P. L. (in press). Self-image, science and math: Does the image of the 'scientist' keep girls and minorities from pursuing science and math? *Sex Roles.*

Mestre, J. P., Gerace, W. J., & Lochhead, J. (1982). The interdependence of language and translational math skills among bilingual Hispanic engineering students. *Journal of Research in Science Teaching, 19,* 1255–1264.

Mestre, J. P., & Robinson, H. (1983). Academic, socioeconomic, and motivational characteristics of Hispanic college students enrolled in technical programs. *Vocational Guidance Quarterly, 31,* 187–194.

Mirandé, A. (1977). The Chicano family: A reanalysis of conflicting views. *Journal of Marriage and the Family, 39,* 747–756.

Montez, P. (1960). *Some difference in factors related to educational achievement of two Mexican-American groups*. San Francisco, CA: R & E Research Associates.

Moore, E. G. J., & Smith, A. W. (1985). Mathematics aptitude: Effects of coursework, household language and ethnic differences. *Urban Education, 20,* 273–294.

National Assessment of Educational Progress. (1979). *Mathematical understanding* (Report No. 09-MA-04). Denver, CO: Education Commission for the States.

National Assessment of Educational Progress. (1980). *Mathematics technical report: Summary volume* (Report No. 09-MA-21). Denver, CO: Educational Commission for the States.

National Center for Educational Statistics. (1980). *The condition of education for Hispanic Americans*. Washington, DC: U.S. Government Printing Office.

Pallas, A. M., & Alexander, K. L. (1983). Sex differences in quantitative SAT performance: New evidence on the differential coursework hypothesis. *American Educational Research Journal, 20,* 165–182.

Pallone, N. J., Hurley, B., & Rickard, F. S. (1973). Further data on key influences of occupational expectation among minority youth. *Journal of Counseling Psychology, 29,* 484–486.

Parelius, A. P. (1981). *Gender differences in past achievement, self-concept, and orientation toward mathematics, science and engineering among college freshman*. Toronto, Canada: American Sociological Association.

Pascarella, E. T., Walberg, H. J., Junker, L. K., & Haertel, G. D. (1981). Continuing motivation in science for early and late adolescents. *American Educational Research Journal, 18,* 439–452.

Powers, S., & Wagner, M. J. (1983). Attributions for success and failure of Hispanic and Anglo high school students. *Journal of Instructional Psychology, 10,* 171–176.

Ramirez, M. (1971). The relationship of acculturation to educational achievement and psychological adjustment in Chicano children and adolescents: A review of the literature. *El Grito, 4,* 21–28.

Ramirez, M. (1977). Recognizing and understanding diversity: Multiculturalism and the Chicano movement in psychology. In M. Ramirez (Ed.), *Chicano psychology* (pp. 243–253). New York: Academic Press.

Ramirez, M., Castañeda, A., & Herold, P. L. (1974). The relationship of acculturation to cognitive styles among Mexican Americans. *Journal of Cross-Cultural Psychology, 5,* 424–433.

Ramirez, M., & Price-Williams, D. R. (1974). Cognitive styles of children of three ethnic groups in the United States. *Journal of Cross-Cultural Psychology, 5,* 212–219.

Rendón, L. I. (1983). *Mathematics education for Hispanic students in the Border College Consortium*. Laredo, TX: Laredo Junior College, Border College Consortium.

Roberge, J. J., & Flexer, B. K. (1981). Re-examination of covariation of field independence, intelligence and achievement. *British Journal of Educational Psychology, 51,* 235–236.

Romo, H. (1985). The Mexican origin population's differing perceptions of their children's schooling. In R. O. de la Garza, F. D. Bean, C. M. Bonjean, R. Romo, & R. Alvarez (Eds.), *The Mexican American experience*. Austin, TX: University of Texas Press.

Satterly, D. J. (1976). Cognitive styles, spatial ability and school achievement. *Journal of Educational Psychology, 68,* 36–42.

Schwartz, A. J. (1971). A comparative study of values and achievement: Mexican-American and Anglo youth. *Sociology of Education, 44,* 438–462.

Sells, L. (1980). The mathematical filter and the education of women and minorities. In L. H. Fox, L. Brody, & D. Tobin (Eds.), *Women and the mathematical mystique*. Baltimore, MD: The Johns Hopkins University.

Serna, D. A. (1980). Issues affecting bilingual education in elementary education. In Salient issues in mathematics education research for minorities. *Proceedings from a National Institute of Education sponsored meeting*. (ERIC Document Reproduction Service No. ED 191 699)

Sherman, J. (1980). Mathematics, spatial visualization and related factors: Changes in girls and boys, grades 8–11. *Journal of Educational Psychology, 72,* 476–482.

Smith, G., & Lantz, A. (1981). Factors affecting the participation of minorities in nonrequired mathematics courses. In G. Smith, A. Lantz, & V. L. Eaton (Eds.), *Factors influencing the choice of non-required mathematics courses* (pp. 32–54). Denver, CO: Denver Research Institute, Social Systems Research and Evaluation Division.

Smith, W. S., & Stroup, K. M. (1978). *Science career explorations for women*. National Science Teachers Association.

TenHouten, W. D., Tzuejen, L., Kendall, F., & Gordon, C. W. (1971). School ethnic composition, social contexts and educational plans of Mexican-American and Anglo high school students. *American Journal of Sociology, 77,* 89–107.

Webster, D. W., & Fretz, B. B. (1978). Asian American, Black and White college students' preference for help-giving sources. *Journal of Counseling Psychology, 25,* 124–130.

Wiley, M. G., & Eskilson, A. (1978). What did you learn in school today? Teachers' perception of causality. *Sociology of Education, 51,* 261–269.

Wise, L. L., Steel, L., & MacDonald, C. (1979). *Origins and career consequences for sex differences in high school mathematics achievement*. Palo Alto, CA: American Institutes for Research in the Behavioral Sciences.

Wittig, M. A., & Peterson, A. C. (Eds.). (1979). *Sex-related differences in cognitive functioning: Developmental issues*. New York: Academic Press.

Zinn, M. B. (1980). Employment and education of Mexican-American women: The interplay of modernity and ethnicity in eight families. *Harvard Educational Review, 50,* 47–62.

Chapter 10

Assumptions and Strategies Guiding Mathematics Problem Solving by Ute Indian Students

William L. Leap
Department of Anthropology
The American University

This chapter explores some of the ways in which one group of American Indian elementary school students goes about solving mathematics word problems. Of particular interest to this discussion are the assumptions about problem solving that guide these students as they approach such tasks, the combinations of logic, inference, and basic arithmetic that help individual students construct answers to particular word problems, and the effects that oral and written language skills have on the organization and the outcome of problem solving under these circumstances.

The "problem-solvers" in question in this chapter come from a group of third-, fourth-, and fifth-grade American Indian students who live on the Uintah and Ouray (or Northern Ute) Indian reservation (northeastern Utah) and attend classes at one of the on-reservation, county-operated elementary schools. Non-Indian students, living on or adjacent to the reservation, also attend this school and are classmates with the Ute students throughout their educational careers. Comparisons between Ute and non-Indian student performance on certain problem-solving tasks will prove to be especially important in the following discussion.

BACKGROUND

My interests in this aspect of Ute student education grows out of work with a larger project, ongoing since 1978, concerned with the maintenance and renewal

161

of fluency in the tribe's ancestral language and with strengthening competence in aspects of the tribe's traditional culture. Language and cultural loss has been a topic of considerable discussion among tribal members for many years. Developing an in-school bilingual education program was one way in which tribal members could respond to these concerns, and, as of the fall of 1980, language arts and content instruction in Ute as well as in English began to be made available to elementary school-aged Ute (and, if interested, non-Indian) students.

In some cases, older tribal members were suspicious of these formalized efforts at Ute language instruction. Some remembered times in their youth when Ute students were punished for speaking the tribal language on school grounds; others felt strongly that Ute should be taught in the homes and communities and not in nontraditional domains. Supporters of bilingual instruction countered these arguments by noting the positive effects that Ute language education could have on Ute student English language ability and, by extension, on their overall level of academic achievement within the English-speaking environment of the public school. And even the most conservative of tribal members found they had to agree that something needs to be done to improve the English language facility and the academic achievement levels of Ute Indian students.

The specifics of the problem here are, unfortunate to report, quite typical of conditions facing tribes, rural Indian communities, and urban Indian enclaves all across the United States and Canada. During their first years of schooling, Ute students do as well, or better than, their non-Indian classmates, as measured by their performance on standardized tests and through other, formal means. By fourth grade, however, achievement levels begin to fall and regular class attendance begins to become a problem. School-leaving becomes a real issue by Grades 8 or 9. In fact, according to the Northern Ute Division of Education, 40% of the Ute students who enter the on-reservation, public junior high school never complete their high school education; of those who are graduated less than 25% go on to college or enter some other career or vocational training program, and less than half of those students will complete their chosen course of study there.[1]

One subject area, which proves to be particularly disquieting for many Ute students regardless of grade level, has clearly observable negative effects on Ute student academic achievement, and contributes substantially to Ute student school-leaving patterns, is mathematics education. As has been found to be the case for Indian students from other tribes (see Moore, 1982, for Navajo; Schindler & Davison, 1985, for Crow; Leap et al., 1982, for comparisons between Oneida and Northern Ute; and the general discussions on this issue in Coombs,

---

[1]Post-secondary education completion rates have improved considerably in recent months, due to the effectiveness of the on-reservation-based, tribally controlled vocational education program and the continuing set of on-reservation courses for college credit sponsored by the tribal Division of Education. School-leaving rates for students attending off-reservation post-secondary programs remain unchanged.

Madison, Kron, Collister, and Anderson, 1955; GAO, 1977; and U.S. Civil Rights Commission, 1973), Ute student performance on in-class mathematics activities and on the mathematics skills portions of standardized tests falls consistently below that of their non-Indian class- and age-mates; evidence of performance difference begins to be seen as early as second or third grade. The performance "gap" widens as higher grade-levels are reached, making mathematics achievement one of the key components determining a Ute student's overall grade point average in junior high and high school.

But the effects of mathematics performance extend into affective domains as well. Ute students who remain in school, for example, tend to avoid enrolling in mathematics courses or in other courses where mathematics plays a significant role in course content. Career choices are often made along similar lines, with Ute students rejecting career options that emphasize the need for quantitative skill and favoring career options where qualitative skills are stressed. As a result, there continue to be virtually *no* members of the Northern Ute Tribe trained in the "hard" sciences, in engineering, in energy-related industrial science, or in business management. This situation forces the tribe (and tribal enterprise) to remain dependent on non-Indians for services in these areas, a situation that has serious implications for economic self-determination as well as for political self-sufficiency.

Improving Ute student mastery over basic mathematic skills was one of the content-related goals that the elementary school-based bilingual education program was designed to address. Before project staff could plan learning activities appropriate to that goal, project staff needed to have a clear sense of the factors promoting poor mathematics performance and mathematics "avoidance" among Ute students. Research toward this end was undertaken during the 1979–1980 academic year at the on-reservation elementary school, using data from a second site (the tribally controlled elementary and secondary school at Oneida reservation, Wisconsin) as well as data from the Ute students' non-Indian classmates to help focus and temper project findings.

Analysis found that any number of factors could be associated with poor mathematics performance for Northern Ute students. Significant items within that inventory include (a) negative teacher attitudes toward the prospect of improving Ute student educational skills; (b) home-school disagreements regarding the roles parents and tribal members should play in the education of Indian students; (c) student acceptance of traditional Ute commitments to personal self-reliance and non-dependence on others for assistance and aid; (d) instances where ways of counting and grouping things that are traditional to Ute culture conflict with counting and grouping principles taken for granted by the school's textbooks and curricula. (See Leap et al., 1982, for a complete discussion of the roles these factors play in Ute student mathematics education.)

According to project findings, some of the factors affecting Ute student mathematics performance could be addressed through innovative, school-based bilingual/bicultural instruction; and as one result of this project, bilingual program staff developed a series of units introducing upper-level elementary school students to basic principles of Ute mathematics and to Ute language mathematics vocabulary. Others of these factors had to be addressed through efforts based outside of the classroom, however. This is one reason why, once the funding for the first 3 years of program operation came to an end, the program director applied for, and received, additional funding to support a 3-year parental training effort. Activities sponsored by the project are designed to integrate family, community, and school more closely than current opportunities allow, in hopes that parents, tribal members, and school authorities will come to see new ways they all can participate, as partners, in the education of Ute Indian students.

One factor that was *not* found to affect the mathematics attainment of Ute Indian students was familiarity with the tribe's ancestral language. Fluency in Ute language showed *no* statistically significant association with poor performance on mathematics-related classroom activities or on the quantitative portions of standardized tests. In fact, research found some evidence to suggest that students who spoke Ute and English performed much more effectively on these tasks, and in other areas of content instruction, than did their monolingual, English-speaking Indian classmates. This observation raised questions about the nature of the language skills that Ute students bring into the classroom, an issue I had explored in considerable detail when working with educational needs in other tribal contexts (see, for example, discussion in Leap, 1977; Leap, 1978, and Wolfram, Christian, Leap, & Potter, 1979). And so, once the "mathematics avoidance" research was completed, research attention shifted to an assessment of the language arts needs of Ute Indian elementary school students.

Research since that time (summer, 1981) has identified a number of areas where the structures of Ute student English differ substantially from the corresponding structures found within standard English, while showing intriguing parallels to details in the grammar of the students' ancestral language. Equally important has been the identification of certain conventions of language use (or pragmatics) among Ute student English speakers, conventions that contrast with the use-related details that speakers of standard English would expect under such circumstances, and again, show parallels to the constraints governing Ute language conversational style.

As might be expected, the points of structural uniqueness—especially, the tense/aspect distinctions, the relationships linking subordinate and independent clauses, and distinctions between singular/plural and animate/inanimate references—were found to have definite effects on the success of Ute student oral and written language performance within the classroom. More significant, however, were the ways that pragmatic factors affected the classroom performance: question-answering strategies, styles for processing information from the printed page, resistance to tasks requiring the use of speculation, personal opinion, or judgment, and preference for discussion in terms of concrete rather than hypothetical description are just a few of the use-related features that show Ute English-standard English contrasts. And now that research is paying closer attention to the ways that the form of a question (e.g., true-false, multiple choice, open-ended probe, and so on) and type of task (e.g., list, describe, compare, analyze, infer) affects the quality of Ute student performance in classroom discussions and on worksheets, examinations, and standardized tests, these contrasts between language structure and language use begin to tell us much about barriers to Indian educational equity at Northern Ute.

These findings also place some of the issues raised during the "mathematics avoidance" study into a new and more interesting light. The presence of a distinctively Ute approach to classroom English raises the possibility that equally distinct types of knowledge structures and skills will be found to underlie other areas of these students' classroom performance and academic achievement. Fortunately, data were gathered during the "mathematics avoidance" study that lend themselves quite neatly to consideration of this question: The fieldworker conducted a set of oral interviews with 18 randomly selected ($N = 56$) fourth- and fifth-grade Ute students, all judged to be low achievers in mathematics by their teachers and school authorities. During those interviews, students were asked to solve a series of mathematics word problems (see Table 10.1) while describing to the fieldworker the approach(es) being used to solve the problem and any difficulties being encountered while they did so. During the original study, comments from these interviews had been used to assess the students' levels of accuracy in problem solving; comparisons had been made with the answers to the problems given by their non-Indian classmates, to identify types of skills-areas that seemed to be particularly challenging to students of Ute background. Results of that analysis are discussed, in part, in Leap et al. (1982, pp. 188–228) and also in Leap (1981). Now, given the findings from the Ute student English study, a reanalysis of the interview data seemed appropriate, this time taking student comments on face value to see what they suggest about (a) the *assumptions* orienting Ute students to mathematical problem solving in general and (b) the *strategies* guiding the students' use of those assumptions when particular mathematics problems have to be solved. Comments on each of these issues are presented in the following sections; Table 10.2 summarizes the issues that are explored.

A NOTE ON METHODOLOGY

The word problems at issue in this analysis had been prepared for use in an earlier study of similarities and contrasts between Northern Ute and non-Indian student cognitive skills (see Witherspoon, 1961, for details). During the first days of fieldwork at the elementary school, all of the students in the third, fourth, and fifth grades were asked to work through the problem-set as part of the in-class

TABLE 10.1

The Mathematics Word Problems Used in Study (adapted from Witherspoon, 1961)

1. Count these crosses with your finger and write the number here .... ____
   X X X X X X X X X
2. If you take away one of these crosses, how many crosses will be left? ____
3. If you take away two more of these crosses, how many will be left? ____
4. If you cut an apple in half, how many pieces will you have? ____
5. Billy had 4 pennies and his father gave him 2 more. How many pennies did he have   ____
   altogether?
6. Tom had 8 marbles and he bought 6 more. How many marbles did he have altogether? ____
7. A boy had 12 newspapers and he sold 5 of them. How many did he have left? ____
8. At 7¢ each, what will 3 oranges cost? ____
9. A milkman had 25 bottles of milk and sold 11 of them. How many bottles did he have   ____
   left?
10. Four boys had 72 pennies which they divided equally among themselves. How many   ____
    pennies did each boy get?
11. If a boy was paid $4 a day for working in a store, how many days would he have to   ____
    work to earn $36?
12. If oranges cost 30¢ a dozen and you buy 3 dozen of them, how much change should   ____
    you get back from $1.00?
13. 36 is two-thirds of what number? ____
14. If 3 pencils cost 5¢, what will it cost you to buy 24 pencils? ____
15. If a taxi charges 20¢ for the first quarter mile and 5¢ for each quarter mile thereafter,   ____
    what will be the fare for a two-mile trip?
16. Jones and Smith start to play cards with $27 each. They agree that at the end of each   ____
    game the winner will get one-third of the money which the loser has in his possession.
    Jones wins three games in a row. How much money does Smith have left at the
    beginning of the fourth game?

assignment during one of their mathematics classes. Comparisons were then made between the Ute students' solutions to these problems and the solutions given by the non-Indian students to the same tasks. These contrasts, along with any recurring response patterns (for example, answers in whole numbers even though the problem calls for an answer in fractions) that could be found within the Ute students' answer sheets, helped the fieldworker select the word problems that the Ute students were asked to discuss during the interviews.

Students were given an unmarked copy of the word problems, a pencil, and ample amounts of scratch paper with which to work during the interviews. No attempt was made to limit the amount of time that students could spend on any word problem they were asked to solve. Each interview was tape-recorded and a verbatim transcript of the discussion was then prepared. Transcripts were divided into problem-centered segments, so that the common themes underlying the students' comments as they worked with particular word problems could be reviewed. Also consulted were the students' notes and calculations made on scratch paper while discussing each problem, as well as the comments on each

TABLE 10.2
Assumptions and Strategies Governing Mathematics Problem Solving
for Northern Ute Elementary School Students

Assumptions
1. Word problems are problems, and problems are supposed to be solved.
2. Solutions to word problems can be found by drawing on personal knowledge and expertise; solutions should not be found by relying on the knowledge and expertise of others.
3. Whatever else their intent, solutions to word problems should always be consistent with the perceived intent of those problems.

Strategies
1. Convert the problem into simple addition or subtraction.
2. Use clues from the text and other sources to develop a generalized, open-ended solution to the problem.
3. Disregard details in the text and work directly with the number values presented by the problem.
4. Assess the truth-value of the conditions outlined in the problem, then develop your answer accordingly.
5. Solve the problem in the manner originally intended.

interview given in the fieldworker's notebooks, to see what perspective these sources of data might shed on those common themes.

ASSUMPTIONS: A COMMON FOUNDATION
FOR MATHEMATICAL PROBLEM SOLVING

To understand the ways in which the Ute students in the sample approached these problem-solving tasks, it is first necessary to look at their understanding of the idea of "problem" and at the demands that a "problem" imposes on them once it has been assigned. Importantly, as comments from the interviews show, Ute students treat word problems as *problems* that need to be given answers. They took this obligation quite literally throughout the interviews. In most cases, once the fieldworker asked a Ute student to work with a particular problem, the student came out with *a* solution (accurate or otherwise) immediately after the request was made. If some part of the task was unclear, the student signaled this fact in very straightforward terms to the fieldworker and then waited for the fieldworker to clarify the matter at hand. Once the necessary clarification was supplied, the student quickly constructed his solution to the problem and offered it to the investigator.

Significantly, and quite unlike the stereotypic picture of the "silent" and "uncooperative" Indian child presented countless times in the literature on American Indian education, no student attempted to avoid outright the field-worker's request that some problem be solved. *No* child sat passively and non-responsively during interview activities, and only under the most specific of circumstances (see comments below) did a Ute student use "I don't know" as a way of coping with the task at hand.

Worth examining in this light are the verbal clues used by the Ute students to signal that some portion of a word problem is not entirely clear. The message in these clues is always quite direct and to the point, as the following discussions of problem 8 show:

F: Okay, let's try number eight. Do you understand that one?
U1: No (silence).
F: What that front part means is that one orange costs seven cents.
U1: Oh, twenty, twenty cents.

* * * * *

F: Okay, another way of saying this is that one orange costs seven cents, so how much will three oranges cost? You can use paper if you want to.
U2: Now what do I do?
F: If you have to pay seven cents for one orange, how much will you have to pay for three oranges?
U2: Is this a plus?
F: This is a multiply. Can you do that?
U2: Uh, twenty cents or no?

* * * * *

F: Do you know what to do with this one (problem 8)?
U3: (reads) "The boy," um, seven cents each, what would three oranges cost?" Um, Um, that's, um, that would be, um, that one was hard for me.
F: Do you have any idea where you would start figuring that one out. If you had—
U3: (interrupts) Seven cents each would it cost, three oranges cost, um.
F: Okay, let's see, if you had, what does one orange cost?
U3: I don't know.
F: It costs seven cents, right?
U3: Um, uh-huh.
F: So one orange costs seven cents, how much would three oranges cost?
U3: Twenty-one cents (said without hesitation).

Notice, however, that the students do *not* indicate what sort of difficulty they are having with the given task through their responses. Their statements signal a state of difficulty, but leave it to the listener (in this case, the fieldworker) to infer the nature of the confusion from context and other details, and then to take the lead in resolving the matter on the students' behalf. In other words, even though Ute students will respond positively to external assistance when offered to them by some outside party, Ute students appear reluctant to solicit such assistance for themselves. Apparently, the obligations of problem solving need to be met by relying on one's own skills and abilities, not on the skills and abilities of others.[2]

A third component of the Ute students' assumptions about problem solving also needs to be mentioned: Students assume that each word problem will have a particular issue, focus, or concern and, therefore, answers given to any word problem must in some visible way address that issue, focus, or concern. As comments in specific interviews suggest, Ute students work with this assumption in one of two ways. In some instances, they construct their answer so it will highlight the point of reference they see being emphasized in the original questions. Note, in the following interview segment, how student U3 recognizes change received, not the price of purchasing fruit, as the issue to be addressed in her calculations; her solution to the problem and discussion of that solution are phrased accordingly:

> F: Okay, now (reads) "if oranges cost thirty cents a dozen and you bought three dozen of them, how much change should you get back from a dollar?"
> U3: Uh, okay, thirty cents for a dozen?
> F: Uh-huh.
> U3: A dozen, okay, you give him a dollar, that's a minus, that's a minus it, that means, zero, twenty cents back?

In other instances, Ute students seem more concerned with emphasizing the mathematical, rather than referential, aspects of the assigned word problem in their answers. Student U4's treatment of problem 8 (just cited) is one example of this preference; here, even though the description of her calculations appears to have little to do with the answer she obtained, the wording of the description is totally consistent with the mathematical focus she associates with the problem.

> F: Okay, can you figure that one out?
> U4: (reads) "Eight cents each, what will the oranges cost?"
> F: (reads) "At seven cents each, what will three oranges cost?" That means, one orange costs seven cents, so how much will it cost you to buy three?
> U4: Thirteen and (writes on scratch paper).
> F: (reads from scratch paper) Twenty cents. How did you figure that?

---

[2]This constraint may apply more directly to interactions with adults than it does to interactions with other students since, as is common in many such classrooms, Ute students often attempt to discuss seatwork assignments (and, on occasion examination questions) with their classmates. Even then, by my observation, direct requests for assistance are infrequent and, if offered, are always carefully worded to focus on the other student's answer ("What did you get for problem 6?") but not on the procedures or techniques used to arrive at that solution ("How do you answer number 6?").

U4: Mmm. Let's see. There's seven, and six is twelve, that's thirteen, Mmm. Twelve, seven and seven make fourteen and you add them.

Student U4 made a carefully selected use of a mathematical principle (''and you add them'') to help validate her answer to this problem. Student U1 uses a similar mathematically related reference (''times tables'') as part of the explanation for his answer to problem 10. Note the use of the ''non-directive'' verbal clues to show the need for further clarification of the terms of the problem during the first portion of this discussion.

F: Okay, now what about problem ten. Do you remember how you did that one?
U1: (reads) ''Four boys . . .''
F: Uh-huh.
U1: Um.
F: (reads) ''Four boys had seventy-two pennies which they divided equally among themselves.''
U1: (interrupts) I don't think I did that one.
F: Okay, how would you, if you had to do that one, how would you do it?
U1: Right now?
F: Uh-huh.
U1: (pause) There was thirteen, thirteen pennies.
F: You think each one got thirteen pennies?
U1: Yep.
F: Okay, how did you figure that out?
U1: I don't know. I just remember times tables.

The point is, whichever the option selected, the Ute students' solutions to mathematics word problems are kept consistent with the intent of each problem as they perceive it. There are instances, of course, where students' perceptions of the intent of a word problem may differ substantially from the classroom teacher's understanding of the purpose of the problem. Several such instances, along with the effects these differences in pe-ception had on success in student problem solving are discussed in the following sections of this chapter.

STRATEGIES: A VARIETY OF PROBLEM-SOLVING OPTIONS

Ute students may begin their problem-solving activities by working with a common set of assumptions. However, the specifics of their problem solving and the answers that emerge once problem solving is completed are in no sense going to be uniform across students. Much of this variability can be traced to the range of problem-solving strategies Ute students employ under such circumstances. At least four strategies can be identified within the student interviews, each having its own visible impact on the success of the student's problem-solving efforts.[3]

The first of these strategies leads Ute students to convert the arithmetical operation originally outlined in the word problem into a simple addition (or, but less frequently, simple subtraction) task; to use the students' own terminology here, either you "plus" it or you "take away," depending on the details presented in the problem. Consider student U4's response to problem 8:

> U4: Mmm. Let's see. There's seven, and six is twelve, that's thirteen. Mmm. Twelve, seven and seven make fourteen and you add them.

This illustrates one way in which "conversion" into addition can be introduced into the problem-solving process. A more successful use of this strategy (where accuracy of the answer is concerned, at least) can be found in student U3's response to problem 11:

> F: Okay, let's try this one. (reads) "If a boy was paid four dollars a day for working in a store . . ."
> U3: (reads) ". . . how many days will he have to work in every thirty-six dollars." Mmm.
> F: ". . . to earn thirty-six dollars."
> U3: Okay, he gets four dollars and times thirty-six, and he has to, um, okay that's eight, twelve, sixteen, twenty, twenty-four, twenty-eight, thirty-two, thirty-six, okay, that would be the end. That's um, one, two, three, four, five, six, seven, eight, nine—nine days.

Conversion from multiplication to addition appears to have been a useful problem-solving technique under these circumstances. Use of this strategy does make high per-problem demands on the student's time, something that could be serious during an end-of-quarter examination or on a standardized achievement test. In some cases during the interviews, use of the strategy seems to have increased the possibility that the student would make an error during the calculation. Such turned out to be the case during student U5's work with problem 10:

> F: How many fours in seven?
> U5: One.
> F: And how many left over?
> U5: One?

---

[3]A fifth strategy leads Ute students to solve word problems in the manner originally intended. This discussion is concerned with alternatives to those "expected" responses; in most cases, reasons why alternative strategies might be called into play are also explored.

F: No.

U5: Two? Three?

F: Yeah, okay. You put a three in front of that two, what do you have? Thirty-two, right?

U5: Yep.

F: How many fours in thirty-two?

U5: That's, um, four, no. Let's see. Thirty-two. Eight, twelve, sixteen, twenty, twenty-three, twenty-seven—

F: (interrupts) That's twenty-four.

U5: Twenty-four. (Silence).

F: So, twenty-eight.

U5: Twenty-eight, thirty-two. Nine.

In this case, once the fieldworker pointed out the error and showed how the addition sequence needed to be altered, the student was quickly able to correct the sequence and come up with the correct answer. In other examples, the students appear to concentrate so forcefully on completing the "converted" calculation that the fieldworker's attempts to provide correction is almost completely ignored:

F: Can you figure out how to do that one (problem 10)?

U6: Seventy-two pennies, seventy-two divided by, um (starts to count in a low voice)—

F: . . . you started to say "seventy-two divided by . . . ." What was that going to be?

U6: (still counting in a low voice)

F: Seventy-two divided by?

U6: Four.

F: Uh-huh.

U6: (begins to mark on the scratch paper) Four into four is . . .

F: (corrects the set-up of the division problem)

U6: Four divided by seventy-two equals . . . four, eight, twelve—

F: (interrupts) Do you know how to do long division?

U6: —sixteen—

F: Okay.

U6: Twenty. (pause) Twenty cents each.

F: Okay, that is pretty close.

U6: Twenty-four.

F: No. How did you get that twenty?

U6: I don't know. (Erases part of her calculation and begins the process a second time.)

Certainly, converting the arithmetic tasks outlined in a problem from multiplication or division into simple addition (or subtraction) reflects the student's awareness of the similarities underlying these operations. Even so, as these examples suggest, awareness of those similarities does not prevent other factors from interfering with the accuracy of the answers that this strategy creates.

A second problem-solving strategy also leads Ute students to re-define the mathematics within the word problem. This time, however, the changes that are made allow the student to by-pass the need for precise calculation, replacing it with a more open-ended, unstructured, and generalized response to the conditions at hand. Student U1's discussion *of* problem 8 (cited above) showed one of the ways in which "generalization" was used as a problem-solving strategy during these interviews. For a second example, consider student U7's treatment of the same problem:

> F: Let's see, number eight. Another way of saying one orange costs seven cents, so how much will three oranges cost. Can you do that one?
> U7: Is it twenty-eight cents?
> F: That is close. Try again. If they are seven cents each and there are three of them, how much will it cost in all?
> U7: Twenty cents.

Note the patterning underlying student U7's responses to this problem: The first response ("twenty-eight cents") seems almost arbitrary in its numerical focus, given the other details in the problem. The fieldworker's rejection of this response placed the student under an even greater obligation to come up with an acceptable answer to this word problem. Constructing a response in cautious terms, so that the student would not be forced to commit herself to some specific (and, once again, possibly inaccurate) number value, seems an especially useful strategy under such circumstances. "Twenty cents" is not a correct answer to problem 8, but it is not an inappropriate answer when viewed in terms of a Ute student's *total* reaction to this problem.

Use of a third problem-solving strategy (namely, disregard the words in the problem and work exclusively with the number values) can be seen in student U3's approach to problem 12:

> F: Okay, now (reads) "If oranges cost thirty cents a dozen and you bought three dozen of them, how much change should you get back from a dollar?"
> U3: Uh, okay, thirty cents for a dozen?
> F: Uh-huh.
> U3: A dozen, okay, you give him a dollar, that's a minus, that's a minus it, that means zero, twenty cents back?

Even though the student appears to have recognized that this was a subtraction problem ("That's a minus, that's a minus it") the problem does not appear to have been solved in the manner originally intended. The answer does make sense, however, if, once the focus of the problem has been established (as required under the terms of assumption 3, as previously discussed), the wording of the text is then by-passed and the analysis centers exclusively on the number values in the problem that relate directly to that focus. Elsewhere in the corpus, Ute students are found to subtract smaller number values from larger number values, regardless of the meanings assigned to those values by the problem. Such appears to have been the case in this instance: After determining that this problem has to do with making change, student U3 scanned the text to find the number values associated with that focus ("thirty cents," "a (one) dollar"), then subtracted the smaller value ("one") from the larger value (the "three" in "thirty"), casting the results of that subtraction in terms that will also be consistent with the problem's identified focus.

Student U8's treatment of problem 11 contains another instance of this "numbers-only" problem-solving strategy.

> F: Okay, can you figure out that one (problem 11)?
> U8: Four dollars (silence), thirty-six (silence), dollar.
> F: Uh-huh.
> U8: (working sounds, counts out loud)
> F: (interrupts) Okay, you added them.
> U8: (silence)
> F: Look at it this way. If you have, let's see, he gets four dollars a day, right?
> U8: Uh-huh.
> F: How many dollars is he going to have in two days?
> U8: Six.

Once again, the significance of this answer can be seen when it is considered as part of a Ute student's *total* reaction to such a problem. Initially, as the student's verbal clues suggested, the meaning of the problem was not entirely clear. And left to his own devices (assumption 2), the student attempted to solve the problem through the most straightforward of strategies: convert the problem to basic addition, and add up the number values as appropriate to the focus of the problem. The fieldworker responded to this move by attempting to refocus the student's attention on the issues raised by the problem:

> F: Look at it this way. If you have, let's see, he gets four dollars a day, right? . . . How many dollars is he going to have in two days?

The student's response is immediate:

> U8: Six.

suggesting (assumption 1) that (this part of) the task has now become manageable. And the student's response makes sense, in the context of the fieldworker's question, once the content-issue raised by the question (how much will be earned over a 2-day period?) is disregarded and basic addition is applied to the two number values ("four" and "two") that are now at issue in this discussion.

Student U2 answered problem 8 in similar terms:

> F: How about number eight? You understand what it is asking you to do?
> U2: (reads silently) It'd be ten cents in all.

Ten cents is not an unreasonable answer to this problem, if analysis treats the task as a "plus" and then proceeds to "plus" the two number values in the text ("seven" and "three") without further reference to the meaning of the text. Importantly, when the fieldworker asked the student to re-think the analysis:

> F: Okay, another way of saying this is that one orange costs seven cents, so how much will three oranges cost? You can use paper if you want to.
> U2: Now what do I do?
> F: If you have to pay seven cents for one orange, how much will you have to pay for three oranges?
> U2: Is this a plus?
> F: It is a multiply. Can you do that?
> U2: Um, twenty, or no?
> F: Twenty is close. Can you figure it out on the paper?
> U8: Twenty-one?
> F: Mm huh. How'd you figure that out?
> U8: Mm.
> F: Did you keep adding to seven to seven? Like that?
> U8: Uh-huh.

The student ultimately arrived at the correct answer for this problem. Here, much as was the case in student U3's discussion of problem 12 and student U8's discussion of problem 11, the combination of assumptions and strategies underlying the student's problem-solving effort is worth noting. An initial clue that more information was required ("Now what do I do?") was followed by a generalized response (strategy 2) to the problem ("Twenty or no?"). When that response was rejected by the fieldworker ("Twenty is close. Can you figure it out on the paper?"), the student shifted his approach yet another time and, by converting the task to simple addition (strategy 1), arrived at the accepted solution. The student's obligation to solve mathematics problems once they are assigned (assumptions 1 and 2, as discussed above) is evident throughout these examples; their use of a *series* of problem-solving strategies, substituting one approach for a second approach until a correct answer (or something suitably close to it) is found, is certainly understandable under such circumstances.

There is one more problem-solving strategy that, although not widely attested within the interview data, occurs frequently enough in the corpus, in classroom discussions, and in the Ute students' written work to make its discussion here seem quite worthwhile. The use of generalized, rather than problem-specific solutions to word problems was noted at an earlier point in this section. In some cases, Ute students will go even further with idea, to come up with solutions to mathematics problems that avoid any need for calculation whatsoever. Consider, for example, student U7's treatment of problem 10:

F: Okay, let's try to figure this one out. This one stumps just about every-body. Just about nobody got that one.
U7: (reads) "Four boys had seventy-two pennies." (reads silently)
F: Can you figure out how you're going to do that
U7: (silence)
F: Let's imagine that you have seventy-two pennies right here in a pile, and there's one boy sitting there, and one boy sitting there, and one boy sitting here and one boy sitting here (points to four places on scratch-paper). What would you do to make certain that everybody got the same number of pennies?
U7: Pass them out until they are all gone.

Not all of the solutions to mathematics problems constructed in terms of this qualitative problem-solving strategy are as practical or as pragmatic as was this suggestion. In other instances, students appear to be reacting more to the amount of reality contained in the terms of the word problem, than they are attempting to develop a functional response to it. During a later point in one of the interviews, a Ute student was asked to determine how much his brother would have to spend on gasoline if he wanted to drive his truck from the reservation to Salt Lake City. Instead of estimating (or generalizing) a response, or attempting to calculate an answer based on the information presented in the request, the student responded quite simply:

U9: My brother does not have a pickup.

At issue in both of these examples is the Ute student's assessment of the "truth value" of the conditions presented by a word problem. Such assessments are a common part of everyday life on this reservation, where persons of all age levels are concerned. Decisions regarding personal conduct as well as innova-tions and changes affecting the tribal membership as a whole are made only if the proposed line of action seems consistent with recognized cultural convention. Proposals that conflict with "the way things are" may be briefly considered, then set aside for action "at some later time," or may simply be given no serious discussion at all.

Viewed in these terms, it is not surprising to find Ute students evaluating the "degree of reality" presented within the text of a mathematics problem and constructing their response accordingly. Ute philosophy takes precedence over non-Indian perspectives in other areas of daily life, so why should Ute philosophy not take precedence over the content of classroom instruction?

LANGUAGE AND CULTURE AS FACTORS IN PROBLEM SOLVING

Thus far in this chapter, the discussion has focused on some of the assumptions and strategies that guide Ute elementary school students' work in mathematical word problem solving. To be complete, however, the discussion also needs to examine some of the factors that make this approach to problem solving such an attractive cognitive strategy for Ute students.

The role played by language-related factors in structuring these assumptions and strategies is of particular interest in this regard, given the parallels that can be drawn between Ute student approaches to mathematical problem solving and their responses to other types of analysis/communication tasks within the classroom.

Consider what happened, for example, when a group of Ute (and non-Indian) fourth graders were asked to read a story and then to discuss events in the story and their reactions to them with the interviewer. The story described the exploits of Slim Green, a small green garden snake. At the beginning of the story, Slim Green is taking a nap and is awakened from his sleep when Rowdy the dog comes running by. In the discussion of story events with the student readers, each student was asked:

Q: If you looked at Slim Green, could you tell if he was sleeping?

Almost all of the 15 Ute fourth graders interviewed answered the question by repeating the observation made on this point in the text: It would be difficult to tell whether Slim Green were awake or asleep, since his eyes were always open. Over a third of these students, when asked *why* (or, in "proper" Ute student English, *how come*) his eyes were open, departed from the text, and offered types of explanations for this condition:

- Cause other things might get him.
- Maybe he couldn't sleep.
- Snakes never sleep.
- Cause his eyes are green.

None of these answers conflicts with the facts of the story, but they certainly do not project an awareness of this portion of the original narrative. Yet awareness of the narrative may not be the issue being addressed through these responses. The students were asked to explain whx Slim Green's eyes were open and each of these statements provides an explanation for this condition. Story content notwithstanding, the students are satisfying the conditions of the given task.

This is the same type of outcome that emerges when Ute students solve mathematics problems in terms of assumption 3: Whatever the numerical details of their answer turned out to be, the students always attempted to acknowledge the qualitative focus of the problem in the wording of their answers. This concern with continuity between question and response, so basic to the Ute students' work in mathematical problem solving (see assumption 3 in Table 10.2, and the previous discussion of that assumption). It is, in addition, a special case of a more general discourse strategy that helps Ute students structure their responses to other types of content-questions as well.

Parallels beyond the domain of mathematical problem solving can also be found within some of the Ute students' problem-solving strategies discussed in the previous section. The priority of "numbers" over "words" suggested under strategy S4 is frequently found during Ute student reading activities, regardless of the subject matter that is being addressed in the reading assignment.

At one point during the "Indian English" study, Ute and non-Indian fourth graders were asked to read a story, the beginning of which is as follows:

Warm spring rains were beginning to melt the snow around the entrance to the den. Sometimes a ray of sunlight beamed in on the bears. One day Major swatted at a sunbeam with his paw and started to follow it out of the den. His mother picked him up, taking his whole head gently in her paws. She put him back in the middle of the nest. It was not yet time to leave the den.

By the middle of April, Major and his sister each weighed almost seven pounds. Their fur was thick and dark and woolly. They romped and wrestled all over the den, tumbling over their patient mother's back.

One morning the mother bear sat up and yawned. She went to the den entrance and looked out. She sniffed the air and then stepped outside, grunting to the cubs to follow her. Major clambered out through the entrance first and his sister followed.

Outside was a strange new world for the cubs. Major blinked his eyes in the bright, dappled sunshine. He looked at the great tree trunks that towered toward the sky. He sniffed a patch of snow in a shady spot and touched it with his paw. It was cold and wet. The breeze whirled some dead leaves in his face. Major whoofed with surprise and scampered back to his mother's side.

That night the bears returned to the den to sleep. Soon they abandoned the den, however, and slept on beds of leaves in thickets.

Then, the students were asked a series of questions, one of which—"How old was Major?"—deliberately probed each student's ability to construct an inference about Major's age based on such facts as:

1. His mother "took his whole head in her paws";
2. Major tried to leave the cave before spring had come;
3. Once he left the cave, "outside was a strange new world for Major and his sister."

Appropriate inferences regarding Major's age, under these circumstances, would include:

- Major is a new-born bear cub.
- Major is less than a year old.
- Major is a baby.

Surprising to note, in the first analysis at least, were the number of instances where Ute students concluded that Major was 7 years old. This response disregards the age-related clues presented in the text and, equally seriously, distorts the meaning of the sentence in which the key word in this analysis ("seven") is found:

By the middle of April, Major and his sister each weighed almost seven pounds.

The point is, "seven" is a comment about Major's weight, not about his age. In fact, the text makes no specific reference to his age, hence the need to construct and answer this question in terms of the points of inference just described.

Apparently, however, some Ute students answered the question in terms of a different set of inferences. This being a question about age, and age usually being described in numerical terms, they searched through the text to locate a number that could then be used to respond to the question; the meaning the text assigned to that number became of secondary concern, given the terms of this inference strategy. So here again, another aspect of these students' approach to mathematical problem solving needs to be seen as a special case of a more general rule of Ute student English discourse.

Notice in addition how something similar to the constraints of assumption 1, namely, that mathematics problems are problems and problems are meant to be solved, can also be found to be operating within student responses to both of these reading-related tasks. Although the students' answers may not always have been consistent with the questions' original intentions, the Ute students did attempt to meet their part of the responsibilities in each discussion by replying in some appropriate form, to each of the questions presented to them. In these cases, as in their work with mathematics word problems, requests for information impose obligations and are not, therefore, to be taken lightly.

None of these parallels are especially surprising. The mathematics-related "knowledge structures" described in the first part of this chapter serve to define the presuppositions Ute students bring to the process of mathematical problem solving; and presuppositional structures, once identified, always turn out to be integral to a speaker's linguistic performance within any number of linguistic domains.

What cannot be overlooked, however, is the fact that, as a group, the elementary school-aged Ute students at issue in this discussion come from homes or family backgrounds where the tribal language, as well as (one or more varieties of) the national language, are commonly spoken. So the possibility of "input" from both language sources (and from other languages as well, depending on home and family context) always need to be addressed, whenever language-related influences on student classroom behavior are to be explored.

Parallels between constraints of Ute language grammar and usage conventions and Ute student approaches to mathematical problem solving can easily be identified. Two of the meanings associated with assumption 2 in this discussion, namely, avoidance of direct questioning and de-emphasis on tightly focused requests for assistance from outsiders, are familiar points of conversational style for Ute speakers of all ages. Ute student use of addition when the word problem calls for multiplication (strategy S1) suggests the fusing of these two concepts under a single lexical item that always happens during ancestral language discussions of western mathematics.

The Ute students' commitment to evaluating the "truth" of a given problem's content and to work with that problem according to the findings of that evaluation offers another example of such parallels. Ute language grammar provides precise mechanisms for indicating whether an event under discussion really happened, might have happened, could have happened, or is merely a topic for speculation and/or fantasy. Similar conventions to distinguish "degrees of reality" can be found in the oral and written English of children and adults from this reservation. In some instances (particularly those associated with conditions-contrary-to-fact), expressions making these distinctions violate the rules of standard English grammar *and* of Ute English as well; such a use of grammatical error seems an especially effective way of indicating the *non*-reality of the issue under discussion in such instances. Ute student assessments of "truth value," and response to word problems in terms of those assessments, merely extends this principle of linguistic classification into another communicative domain.

The interesting thing about Ute language parallels in these (and other) instances is this: Even though all of the students in the interview sample came from Ute language-related home and community backgrounds, none of the students in the sample were fluent speakers of Ute and only two of the students were judged by Ute-speaking classroom aides to be familiar with the Ute language in any more than a casual way. So if features from Ute language tradition are affecting these students' work with problem solving, the details from that tradition have become incorporated into their mathematics-related "knowledge structures" independently of any acquisition of Ute language competence.

The "missing variable" here, as noted in an earlier section of this chapter, turns out to be the variety of English spoken by young people (and adults) on this reservation. "Ute English" (as it has been termed throughout this discussion) includes rules and constraints from Ute as well as English language tradition within its grammar, and the fluent speaker of Ute English maintains, through knowledge of this English variety, a considerable degree of familiarity with his ancestral language *whether he is a speaker of that language or not.* Ute English is the variety of English which most elementary-aged Ute students learn as their first language. It commonly is the students' primary means for communication inside and outside of the classroom; it serves as the students' model for oral language use, and it has definite effects on Ute student written English style as well. More important for present purposes, Ute English is the language in which mathematics is *learned* and (depending on the person providing the instruction), Ute English may also be the language in which mathematics is taught. Viewed in those terms, and recognizing that most Ute students come from homes and communities where other aspects of tribal traditions are still very much a part of daily life, it is no wonder that details from Ute language tradition help structure Ute student approaches to mathematical problem solving, even when the students are not speakers of the tribal language.

THE CLASSROOM AS A PROBLEM-SOLVING ENVIRONMENT

Not to be overlooked here are the effects the elementary school classroom itself has had on the formation and maintenance of this language-related approach to problem solving. Developing a detailed description of the classroom context was one of the objectives of the "mathematics avoidance" study. Research to meet this objective was organized according to Cicourel's (1974) concept of "indefinite triangulation"; that is, information on classroom activities and perceived strengths and weaknesses of those activities was gathered from students, teachers, other school authorities, parents, and significant tribal officials, as well as through a series of systematic observations of day-by-day activities during mathematics instruction. (See Leap et al., 1982, pp. 208–222) for a detailed description of this methodology.) Information from these sources was compared, to identify common themes and concerns running through the commentaries, to locate significant points of contradiction, and to develop explanations that would account for those contradictions.

Since the Ute students were the key individuals in the study, it seemed important to determine how they perceived themselves, as Utes, as elementary school students, and as mathematics learners. To do this, a series of values clarification activities were included in the opening part of the student interviews. Student responses to these activities were then encoded and factor analysis procedures were used to identify the associations between responses that recurred most frequently within the corpus. Primary elements within the Ute students' profile of "self and others" perceptions emerging from this analysis included the following descriptors:

- I am able to get what I need by myself.
- I am able to make up my mind with little trouble.
- It is important to take from nature the things that you need.
- It is important to understand the ways of nature.
- It is not important to tell others what to do.

as well as:

- Students in my mathematics class are not friendly to me.
- Other kids pick on me.
- My friends do not like the teacher.
- No one pays attention to me at home.

and:

- I do not talk in mathematics class.
- My teacher does not help me enough.
- I do not learn much in mathematics class.

Leap et al. (1982) used the term *self-dependence* to summarize the perspective on "self" being suggested in the first set of comments. That is, according to this analysis, Ute students as a group find it appropriate to follow their own initiatives inside and outside of the classroom, to take personal responsibility for the consequences of those actions, to expect that other people will also pursue their own initiatives in terms of their own sense of responsibility, and to assume that other persons will extend the same respect for personal autonomy to them. This is, of course, nothing more than a formal and quantitatively justifiable restatement of the second assumption about the process of mathematical problem solving discussed in the previous section: Ute students solve such problems by relying on their own skills and abilities, not by seeking assistance from other parties.

The remaining comments in this listing, augmented by information from other interviews, from classroom observations, and from other sources, help clarify why such an attitude of self-dependence should come to be so basic to Ute student experiences with classroom-based mathematics learning. Classroom observations during the conditions implied here are not at variance with the realities of mathematics instruction evidenced at this school during the research period. The 1979–1980 academic year found mathematics instruction, along with instruction in almost every other content area, presented to students primarily in terms of individualized, seatwork experiences. Brief introductory comments might first be made to orient students to their assignment, and occasionally group drills on some technical issue were attempted. In most cases, however, learning was designed as a personal experience. Students were expected to work alone at their seats on the given worksheets or on other tasks. Students needing additional help with the assignment were encouraged to seek assistance from the teacher. To do that, the student had to get up from his desk and cross the classroom to the teacher's desk, and wait until she was willing to take a break from paper grading, administrative activity, or whatever other task in which she was engaged; in some instances, this involved a 3- to 5-minute wait, reducing the amount of time that could be devoted to the assigned task, once the student returned to his seat. It was not uncommon to see students visiting with their friends, or otherwise becoming distracted by other activities while on their way to or from the teacher's desk. This further reduced the student's available worktime, added to the noise within the classroom, and made it even more difficult for serious students to concentrate on their assignment.

The teacher collected the worksheets, once the time set aside for the task came to an end. Assignments were always graded, though this usually involved merely noting if the student had completed the assigned number of problems, not checking to see if the answers given to the problems were in fact the correct ones. In some cases, ''graded'' papers were returned to the students; in other instances, papers were shown briefly to each student, then tossed into the wastepaper basket. Students who did not complete the required number of problems were asked to work on the assignment at home that evening; otherwise, homework was not required of students in any of the classes observed during the research period.

Whether homework assignments would be able to make positive contributions to the education of these students is not entirely clear, according to the field data. Home environment and home-based activities were found to play minimal roles in the schooling experiences of Ute students. Teachers admitted that they would like to see things otherwise and argued that ''they had done all they could do'' to get parents more involved in their children's education. They claimed that Ute parents refuse to work with school authorities on *any* educationally related issue; they noted that Ute parents do not become involved with the Parent-Teachers Association or with any other auxiliary/support service activities at the school.

Indian parents and tribal officials agreed that there was great separation between home and school on this reservation but they described the reasons for that separation in considerably different terms. Parents claimed that it is the school's responsibility to provide educational services to Ute children, not the responsibility of the home. Moreover, many Indian parents would not be able to become involved in their children's schooling, even if they wanted to do so, given the deficiencies that most parents see in their own formal education. Teachers and school authorities are the professional educators, parents argued; and teachers and school authorities need to be given every opportunity to do what they have been trained to do, namely educate! Hence from the parents' point of view, participation in school-related functions, assistance when students complete assignments at home, or any other attempt to become an active participant in the work of the school would constitute a highly improper and unwarranted intrusion of family and tribe into the school's proper realm of responsibility.

So an accurate description of the context of mathematics instruction at this school needs to acknowledge the following facts:

- School is something that operates entirely separately from home and tribe
- The school expects its students to function as isolated, distinct entities within the classroom, and organizes its lessons and assignments accordingly;
- Parents distance themselves from their children's school-related learning activities.
- The teacher places herself deliberately on the sidelines, while "learning activities" are taking place within the classroom.
- Students are forced to take the initiative in requesting technical assistance from the teacher, if questions arise during a work period; otherwise students are left to complete their assignments on their own terms, using whatever resources are personally available to them.

So it is no wonder that Ute students approach mathematical problem-solving tasks by reference to the language-related assumptions and strategies described in this chapter. A sense of problem solving that obligates the student to solve any set of mathematics problems assigned to him, encourages him to work those problems in terms of his own knowledge and skills and discourages him from turning to outside sources for assistance while doing so, and then provides him with a set of possible problem-solving strategies, at least one of which can always be made to apply to any word problem, is certainly a reasonable orientation for Ute students to possess under these circumstances.

Unfortunately, as has also been shown in this discussion, use of this set of "knowledge structures" has its liabilities as well as its benefits, where the education of Ute Indian students is concerned. Although use of these assumptions and strategies may seem highly functional within the classroom context at this school, the solutions to word problems they generate are not always accurate or acceptable when evaluated from the point of view of the teacher, of school authorities, or of standardized achievement tests. How this contradiction affects

Ute student interest in mathematics as a school subject can only be conjectured, at the present time. But perhaps it is now clear why the "Indian mathematics problem" continues to be a source of major concern for *all* Indian educators, and why, even when the "problem" is recognized, truly effective remediation strategies have yet to emerge.

ACKNOWLEDGMENTS

The discussion in this chapter constitutes a new analysis of issues originally raised in Leap (1981) and in Leap et al. (1982). Funding for the original study of these issues was provided through grant NIE-G-79-0086 awarded by the National Institute of Education to William L. Leap and Charles McNett, Jr., of the Department of Anthropology at The American University. This grant enabled a year of study of "math avoidance" questions on two Indian reservations; Leap coordinated the fieldwork at Northern Ute, and Laura Laylin, doctoral candidate in TAU's Department of Anthropology, conducted the initial set of interviews with the Ute students in January, 1980. Math-avoidance-related research has continued at Northern Ute since the completion of the original study, as part of a larger analysis of Ute and non-Indian student language-related classroom competencies. Gratitude is expressed to the National Institute of Education; to Milton Greenberg, Provost of the American University, for award of sabbatical leave during the fall, 1984, allowing a second period of intensive research at the elementary school; and to Betty T. Bennett, Dean of TAU's College of Arts and Sciences, for enabling additional research on this theme during the 1985–86 academic year.

Research at Northern Ute would not be possible without the continuing support of the tribe's Division of Education, Forrest Cuch, director; the tribe's Johnson-O'Malley program, Irene Cuch, director; and the Wykoopah bilingual program based at the elementary school, Venita Taveapont, director. The continuing cooperation from the principal, teachers, and students at the elementary school is gratefully acknowledged.

Sincere thanks to Rodney Cocking, Jose Mestre, Venita Taveapont, Brett Williams, and Signithia Fordham for their helpful comments and criticism throughout the preparation of this manuscript.

REFERENCES

Cicourel, A. (1974). Introduction. In A. V. Cicourel, K. H. Jennings, K. C. W. Leiter, R. MacKay, H. Mehan, & D. Roth (Eds.), *Language use and school performance* (pp. 1–17). New York: Academic Press.

Coombs, M., Madison, L., Kron, R. E., Collister, E. G., & Anderson, K. E. (1955). *The Indian child goes to school*. Washington, DC: U.S. Department of Interior, Bureau of Indian Affairs.

General Accounting Office. (1977). *The BIA should do more to help educate Indian students*. Washington, DC: U.S. Government Printing Office.

Leap, W. L. (Ed.). (1977). *Studies in Southwestern Indian English*. San Antonio: Trinity University Press.

Leap, W. L. (1978). American Indian English and its implications for bilingual education. In J.

Alatis (Ed.), *International dimensions of bilingual education* (pp. 657–669). Washington, DC: Georgetown University Press.

Leap, W. L. (1981). Does Indian math (still) exist? In Amy Zaharlick (Ed.), Native languages of the Americas (pp. 196–213). *Journal of the Linguistics Association of the South at Southwest, 4*(whole no. 2).

Leap, W. L., McNett Jr., C., Cantor, J., Baker, R., Laylin, L., & Renker, A. (1982). *Dimensions of math avoidance among American Indian elementary school students* (Final report on NIE grant NIE-G-79-0086). Washington, DC: The American University, Department of Anthropology.

Moore, C. (1982). *The Navajo culture and the learning of mathematics* (Final report on NIE grant NIE-G-81-00690). Flagstaff, AZ: Northern Arizona University School of Education.

Schindler, D., & Davison, D. (1985). *English language mathematics concepts and American Indian learners.* Billings, MT: Eastern Montana College, College of Education.

U.S. Civil Rights Commission. (1973). *The Southwest Indian report.* Washington, DC: U.S. Government Printing Office.

Witherspoon, Y. (1961). *Cultural influences on Ute learning.* Unpublished doctoral dissertation, University of Utah, School of Education.

Wolfram, W., Christian, D., Leap, W., & Potter, L. (1979). *Variability in the English of two Indian communities and its effects on reading and writing* (Final report on NIE grant NIE-G-79-006). Arlington, VA: Center for Applied Linguistics.

Chapter 11

Opportunity to Learn Mathematics in Eighth-Grade Classrooms in the United States:
Some Findings from the Second International Mathematics Study

Kenneth J. Travers
University of Illinois at Urbana-Champaign

INTRODUCTION

The low mathematics achievement of language minority (LM) students in United States schools has been well documented. The 1985 *Condition of Education* report of the National Center for Education Statistics notes that the basic skills and problem-solving competencies of blacks and Hispanics are much lower than those of white children (p. 59). Not surprisingly, this problem is reflected at the high school level in other ways, such as disproportionately high dropout rates for LM students. In 1980, the *Condition of Education* identified Hispanics as having the highest dropout rate for high school sophomores of all racial/ethnic groups (the overall rate was 14.4%; the rate for Hispanics was 19.1% and for white, non-Hispanics 13.0%) (p. 210).

Although many factors, such as cultural or linguistic, could be adduced to help account for the difficulties that LM students have with secondary school mathematics, this chapter explores a dimension of schooling that has received relatively little attention in the literature—that of the limited opportunities to learn mathematics that are made available to relatively large proportions of students in school, especially LM students. This exploration is made possible by a model used in the Second International Mathematics Study (SIMS), a comprehensive survey of the teaching and learning of mathematics in some 20 countries around the world. This study, carried out by the !nternational Association for the Evaluation of Educational Achievement, gathered data from national samples of 13-year-old mathematics students (8th grade in the United States) and their teachers in each country. (A parallel study also took place at the 12th grade, but is not treated here.)[1] In the United States, some 280 mathematics classes across the country, together with their teachers, were involved.

The methodology of the study provides a rich context in which to view student achievement data. Data are provided on the content of the mathematics curriculum as well as the extent to which this content has been taught to students. The findings reported in this chapter indicate that teacher coverage of mathematical content (called "opportunity to learn") is a powerful variable in accounting for student achievement. Striking evidence reveals that, to a large extent, low mathematics achievement may be explained by the fact that students have not been taught the content being tested.

Another salient finding is that of the high variability of the opportunity-to-learn measure within the United States as compared, say, with Japan. That is to say, for several of the core topics in the eighth-grade mathematics curriculum there are very large differences between classes in the extent to which they have been taught this subject matter. Therefore, much of the low achievement may be attributed to disparities across mathematics classes in opportunity to learn mathematics.

THE SIMS MODEL

In the Second International Mathematics Study, the mathematics curriculum is viewed as taking on different embodiments at each of three levels (see Fig. 11.1). First, at the educational system level (the school district, the educational region, the state, or the country), there is a body of mathematics that students are expected to learn in a certain grade. This mathematics comprises goals and traditions set by the community of mathematicians or educators that help shape the character of the curriculum. This collection of perceived outcomes, together with course outlines, syllabi, and textbooks, delineates the *intended* curriculum. It is what one would be told, for example, if one were to ask an official of the department of education in country (or state) X, "What do you do about ratio and proportion, or the Pythagorean theorem?"

The second level has to do with the classroom where the curriculum becomes *implemented* or taught by the teacher. As Bloom (1974) has observed, "beautiful curriculum plans have little relevance for education unless they are translated into what happens in the classrooms of the nation or the community" (p. 414).

The third component involves student *attainment,* and concerns the question, What aspects of the curriculum intended by the department of education or school district and taught by the teacher are learned by the student? Or, phrased in a different way, to what extent are student achievement and attitudes ac-

---

[1]Details on SIMS are available in Travers (1985).

| Area of research interest | Example of research methodology |

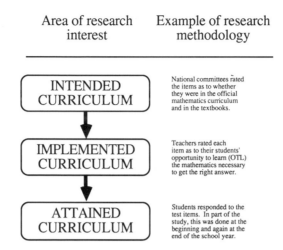

| INTENDED CURRICULUM | National committees rated the items as to whether they were in the official mathematics curriculum and in the textbooks. |

| IMPLEMENTED CURRICULUM | Teachers rated each item as to their students' opportunity to learn (OTL) the mathematics necessary to get the right answer. |

| ATTAINED CURRICULUM | Students responded to the test items. In part of the study, this was done at the beginning and again at the end of the school year. |

FIG. 11.1.   Overview of the IEA Second International Study of Mathematics.

counted for by curricular intentions and by classroom instruction? Since the SIMS model provided for a pretest of achievement at the beginning of the school year as well as an end-of-year test, many countries, including the United States, also have data on how much mathematics was learned *during the school year.*

THE INTENDED CURRICULUM

Data from the Second International Mathematics Study point to the dependence that United States teachers have on textbooks. Figure 11.2 shows that the textbook is by far the most common resource for eighth-grade mathematics teachers, with about 90% of them indicating such usage. Information from analyses of commonly used textbooks, as well as the informed judgment of the U.S. National Mathematics Committee directing the study, provided a profile of importance of various topics in eighth-grade mathematics in the United States. This profile was produced by obtaining from the committee their ratings of the appropriateness for eighth-grade mathematics of each of the items on the international test (Fig. 11.3 gives the results). For example, 100% of the test items on arithmetic, statistics, and measurement were judged to be appropriate at the eighth-grade level, whereas 67% of the algebra and only 57% of the geometry items were so judged. That is to say, in the opinion of the U.S. National Mathematics Committee, the topics of arithmetic, statistics, and measurement as presented on the international test are very important for eighth-grade mathematics but those of algebra and geometry have somewhat less importance. (Other SIMS data tell us that in Japan algebra takes on a great deal of importance for this population, and in France much emphasis is placed on geometry.)

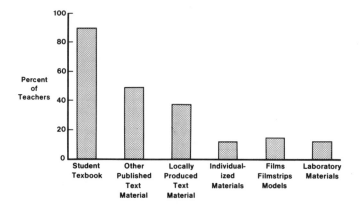

FIG. 11.2. Percentage of teachers using various instructional resources in eighth-grade mathematics classes. (Unless otherwise noted, data are from various reports of the Second International Mathematics Study, Champaign, IL 61820.)

THE IMPLEMENTED CURRICULUM

We now move to the level of the classroom and explore the extent to which topics of arithmetic, algebra, and so on were taught to the eighth-grade classes that were tested. Such information was obtained from the SIMS teacher questionnaire. Each teacher was requested to answer this question for the 180 items on the international test:

Sample test item (for students):
$150 is divided in the ratio of 2 to 3.
The smaller of the two amounts is:
A. $30 B. $50 C. $60 D. $90 E. $120

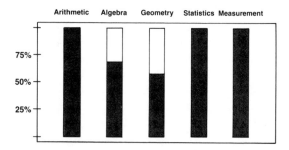

FIG. 11.3. Relative importance of various topics for eighth-grade mathematics in the United States (maximum rating is 100).

Questions for teacher:
1. During this school year, did you teach or review the mathematics needed to answer the item correctly?
A. No   B. Yes
2. If, in this school year, you did not teach or review this mathematics, was it because:
A. It has been taught prior to this school year
B. It will be taught later (this year or later)
C. It is not in the school curriculum at all
D. For other reasons

The resulting "opportunity-to-learn" data are displayed in the box-plots in Fig. 11.4. Box-plots are very informative because they display not only average values (such as for arithmetic or algebra) but also how much variation there is in the data. Leaving aside for the moment the question of variability and focusing on average values (which in this case are medians), we see in Fig. 11.4 five boxes. Across the middle of each box is a horizontal line. This line marks the median value (or middle score) for that particular measure. Take, for example, the opportunity-to-learn data for arithmetic. The middle line of that box is about 85%. So, the graph tells us one half of the U.S. eighth-grade mathematics classes were taught 85% of the arithmetic items on the international test. A different picture is shown for geometry. Here the median value is about 45%. So one half of the classes were taught the content of only 45% of the geometry items.

Overall, the opportunity-to-learn data tell us that across the country, teacher coverage of arithmetic and measurement was quite high, with statistics receiving almost as much attention. This corresponds to the information presented in Fig. 11.3 on the intended curriculum, which shows these three topics as having highest importance in the U.S. eighth-grade curriculum. The remaining topics, algebra and geometry, receive less coverage, with median values of about 74% and 45%, respectively. Again, this pattern corresponds to that of the intended curriculum in Fig. 11.3.

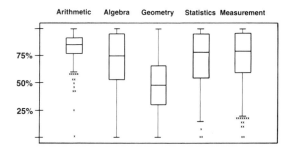

FIG. 11.4. Opportunity to learn various topics in eighth-grade mathematics classes in the United States.

An important characteristic of teacher coverage is that of content variation. That is, we wish to explore the extent to which eighth-grade mathematics classes differ in teacher coverage of arithmetic, algebra, and so on. In the box-plot, the box encompasses the middle 50% of the classes. So, for arithmetic, the middle 50% of the classes were taught between 80% and 90% of the items on the international test. The "whiskers" (the lines protruding from each end of the box) include the middle 95% of the classes. The dots indicate teachers (classes) that are outliers. Compare, for example, the topics of arithmetic and measurement. Although the median amount of mathematics taught is similar, a large proportion of classes was taught virtually no measurement (fewer than 25% of the items). But only two classes were reported as having been taught very few of the arithmetic items. These box-plots show that although there is relatively little variation in teacher coverage of arithmetic (since the box is fairly narrow), coverage of the remaining topics varies a great deal. Now consider the case of algebra. The box-and-whisker plot indicates that opportunity to learn algebra ranges all the way from 0% to 100%. That is to say, there were some classes around the United States that were taught *none* of the algebra on the international test whereas other classes were taught *all* of the algebra on the test. (Remember that the "whiskers" on the graph include the middle 95% of the classes.)

This large variation in opportunity to learn the mathematical content that was judged to be an important part of the intended curriculum for eighth-grade children in the United States represents dramatic inequalities in access to education. For example, the large variation for algebra is a striking contrast to the Japanese setting, as shown in Fig. 11.5. As seen on the left the box-plots are narrow for both arithmetic and algebra in Japan. This represents small dif-

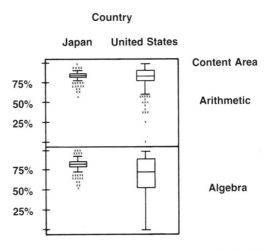

FIG. 11.5. Opportunity to learn arithmetic and algebra in the United States (eighth grade) and Japan (seventh grade)

ferences in class-to-class coverage of these topics across that country. Apparently, in Japan, children have relatively equal access to the study of these topics.

THE DIFFERENTIATED EIGHTH-GRADE MATHEMATICS CURRICULUM IN THE UNITED STATES

How can this great variation in opportunity to learn eighth-grade mathematics be accounted for? A natural place to look for variation is at the school level. We accept the fact that schools do differ in their curricular offerings, and in the extent of those offerings (availability of elective courses, and so on). However, an analysis of the SIMS data by Kifer (1984) indicates that from an international point of view, between-school differences in opportunity to learn mathematics are less in the United States than in several other countries.

The results of one of the Kifer analyses are in Fig. 11.6. In this analysis, a components of variance was done on the pretest scores of the classes in the national samples for the indicated countries in order to determine sources of variation in the data. These sources were identified as due to students, to classrooms within schools, and to schools within each country.

Kifer's analysis shows some intriguing between-country differences. We see, for example, that in Japan, there is very little variation between schools/classrooms in mathematics achievement.[2] (It amounts to about 10% of the total variation.) The remaining 90% is due to differences between students. Apparently, in Japan, schools and classrooms differ very little from each other, and as Fig. 11.5 showed, these classes are taught a great deal of mathematics. It is evident from Fig. 11.6, however, that the school component of variance for the United States is among the smallest of any country. In British Columbia, supplementary information indicates that tracking or ability grouping at the class level is against Ministry of Education policy. We therefore conclude that between-class variation in that Canadian province is small (classes within schools differ little from each other on their pretest scores) and that the majority of the variation is due to schools. In the case of Belgium (Flemish), the bulk of the variance component represented in the cross-hatched region is for schools, since by policy, children are tracked by ability into different kinds of schools.

In the United States there is relatively little between-school variation in opportunity to learn mathematics, so we need to look at the class level data. Here, as again shown in Fig. 11.6, we find that between-class differences are what Kifer describes as "whopping." Only in New Zealand is there a between-classroom variance component of comparable magnitude.

---

[2]Due to sampling constraints, it was not possible to disentagle classroom effects from school effects in four of the countries (Thailand, Japan, Canada [British Columbia], and Belgium [Flemish]).

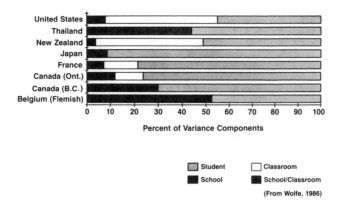

FIG. 11.6. Proportion of variance components in pretest score due to school, classroom, and student

What is the nature of this between-class variation? On the basis of information provided by the teacher, and the textbook used, the U.S. National Mathematics Committee identified four class types (the number of each is indicated in parentheses): Remedial (28), Regular (165), Enriched (31), and Algebra (35). In Fig. 11.7, the opportunity to learn mathematics for each of these four class types in the United States is shown. It is clear from Fig. 11.7 that substantial differences in opportunity to learn mathematics are imposed on students in these four class

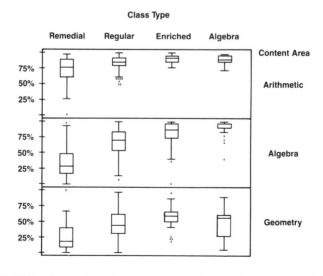

FIG. 11.7. Opportunity to learn mathematics: class type by content area (eighth grade, United States).

types. It can be stated unequivocally that students are following four distinct implemented curricula.

Students in Remedial classes, for example, are provided a curriculum that is predominantly arithmetic. This consists of material to which they had been exposed in previous years and characteristically have been unsuccessful in learning. These Remedial classes, on the other hand, have been taught very little algebra or geometry. At the other end of the spectrum, the Algebra students receive a fairly full dose of high school freshman algebra, but little geometry. Similar comments could be made about the implemented curricula for students in the Regular and Enriched eighth-grade classes.

What are the students in these four class types like? Kifer (1984) did a further analysis of the characteristics of these students, beginning with an assessment of their arithmetic skills at the beginning of the eighth grade. He was interested in the extent to which participation in Algebra classes was merely reflecting differences in prior achievement in arithmetic, or whether other factors appeared to be at work. His results are fascinating. He found, for example, that although arithmetic achievement works rather well as a sorting mechanism for Algebra classes, almost two thirds of the students whose pretest arithmetic scores were in the top quarter were *not* in Algebra. Kifer observed, "although those who get into algebra classes are those who score best on the presumed prerequisite, arithmetic, far more students with equal scores do not have similar access to algebra" (p. 18).

Even more relevant to this volume were Kifer's findings with respect to ethnicity. He found that among the most able students, access to algebra was independent of ethnic background. However, for students of more modest ability (again, as measured by the arithmetic pretest), white students had a much better chance to get into Algebra classes than did their nonwhite counterparts (including LM students).

THE ATTAINED CURRICULUM

At the level of the attained curriculum, one is interested in how much mathematics is learned and how the students feel about mathematics. In Fig. 11.8, achievement in algebra in selected SIMS countries is plotted against their opportunity-to-learn scores. We see, for example, that Japan and France had high achievement in algebra and Sweden and Luxembourg had low scores. The United States scored at about the international median in algebra. What is important, however, is to note that the opportunity-to-learn data for algebra show about the same pattern as the corresponding achievement scores. Japan and France have high opportunity-to-learn scores but those for Sweden and Luxembourg are low. For the United States, opportunity to learn algebra is at about the international average.

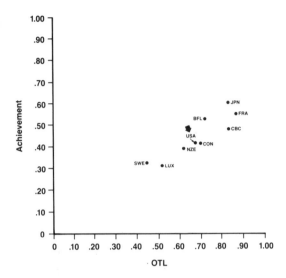

FIG. 11.8.   Achievement vs. Opportunity to learn (OTL) for algebra: (Population A).

Key: BFL—Belgium (Flemish)
      CBC—Canada (British Columbia)
      CON—Canada (Ontario)
      FRA—France
      JPN—Japan
      LUX—Luxembourg
      NZE—New Zealand
      SWE—Sweden
      USA—United States

Let us now look at achievement in algebra in the United States in light of data on opportunity to learn the topic. We have already reported vast differences in opportunity to learn between the four class types in eighth-grade mathematics. Figure 11.9 reports opportunity to learn and corresponding achievement in algebra for these class types. Three kinds of data are shown: opportunity to learn, achievement and the beginning of eighth grade, and achievement at the end of eighth grade. Opportunity to learn algebra is presented by the hollow bar. Hence, for example, the Remedial classes were taught about 35% of the algebra items and the Regular classes were taught about 65% of the algebra items. With respect to student achievement we see that the students in the Remedial classes answered about 20% of the algebra items correctly at the beginning of eighth grade, and about 25% at the end of the school year. (Because the test items were multiple-choice, with five options, a score of 20% is expected by chance!)

The overall pattern of performance shown in Fig. 11.9 is very revealing. First, as already noted, students in the four class types have increasing opportunity to learn algebra as one goes from Remedial, to Regular, to Enriched, to Algebra classes. Second, student achievement increases in a pattern corresponding to increased opportunity to learn. Third, and perhaps most important, the increase in student achievement is strictly step-wise; in other words, students in Remedial classes at the *end* of eighth grade achieve less than students in Regular classes at the *beginning* of eighth grade, and so on for each class type. Clearly, as this pattern continues, the advantage of the student in Algebra classes over those in other classes is dramatic. Indeed, it is doubtful whether the students in the Remedial classes could ever catch up to their counterparts in the more challenging classes.

Another fruitful direction for analysis is that of unusual or unexpected response patterns on test item data. What we report here is based on work by Harnisch and Linn (1981) on the Illinois Inventory of Educational Progress (IIEP) data. These authors point out that unique background experiences may make items that are very easy for most students very difficult for others. Indices have been developed that measure the degree to which the response pattern for an individual is unusual. As these researchers point out, such information can be used "(to) identify individuals for whom the standard interpretation of the test score is misleading, or identify groups with atypical instructional and/or experiential histories that alter the relative difficulty ordering of the items. Furthermore, the items that contribute most to high values on an index for particular

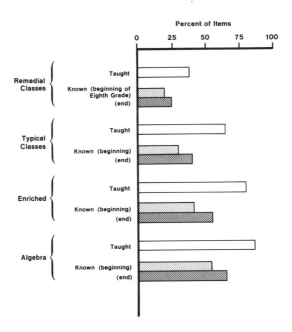

FIG. 11.9. *ALGEBRA* achievement of eighth-grade students by class type.

TABLE 11.1
Weighted Standardized Mean Residuals of Within School Item
Difficulties by Content Category

Content Category	School						
	1	2	3	4	5	6	7
Calculation	−.07	−1.34	−.13	.11	.56	−.58	1.32
Definitions	1.27	.39	−.24	.22	.78	.00	−.32
Numeration	.00	−.66	−2.13	.78	.29	−.81	−1.64
Story problems (general)	−.03	−.79	−.57	.63	−.76	.19	−.60
Story problems (money)	−.10	2.60	1.51	−1.91	−1.29	.01	−.30
Metric systems	−.10	−.56	.88	.56	1.08	2.28	2.20
Figures (fractions)	−2.29	4.16	2.94	−1.08	−1.64	−1.89	4.33
Unclassified	.93	−.85	.07	−1.44	1.61	.28	−.41

subgroups could be identified and judgments made regarding the appropriateness of the content for those subgroups'' (p. 133). Various indices, for example, Sato's caution index and Donlon and Fischer's personal biserial, have been used to identify such patterns.

Harnisch and Linn investigated in some detail the IIEP mathematics data from schools with large caution indices and analyzed student performance on several categories of mathematical content. Table 11.1 (Table 5 from Harnisch & Linn, 1981) shows student performance in various content categories according to content and format. A residual score indicates a difference between expected and observed proportion correct. An entry of 2.0 in absolute value indicates that the items in that category were much easier (+) or much harder (-) for students than expected based on their overall performance and the relative difficulty of the items in the statewide sample. For example, students in School 1 did considerably less well than expected on items involving the figural representation of fractions, and Schools 2,3, and 7 did much better than expected. Harnisch and Linn observe: ''This (finding) suggests the hypothesis that the use of figures to represent fractions may be quite common in Schools 2,3, and 7, but rare in School 1'' (p. 145).

With a view to exploring aspects of the instructional context for LM students, one could propose such an analysis to identify students whose learning of mathematical concepts is facilitated by the use of figures instead of, or in addition to, the use of symbols and words.

SUMMARY

In this chapter, some examples of the kinds of analyses that can be carried out to help explain achievement patterns in junior high school mathematics are demonstrated. In particular, data are presented that indicate dramatic differences in the quality of the instructional context in which students are placed. It is important to examine the extent to which language minority students are subject to such unequal opportunities for quality instruction.

Once such conditions are identified, steps may be taken to remedy deficiencies. The placement of LM students in tracks that greatly reduce opportunities to advance in mathematics is one such apparent practice that warrants further scrutiny.

REFERENCES

Bloom, B. S. (1974). Implications of the IEA Studies for curriculum and instruction. *School Review, 82,* 413–435.

Harnisch, D. L., & Linn, R. L. (1981). Analysis of item response patterns: Questionable test data and dissimilar curriculum practices. *Journal of Educational Measurement. 18*(3), 133–146.

Kifer, E. (1984). *Issues and implication of differentiated curriculum in the eighth grade.* Lexington, KY: University of Kentucky. (Mimeo)

National Center for Education Statistics. (1985). *The condition of education* (A Statistical Report). Washington, DC: U.S. Government Printing Office.

Travers, K. J. (1979). The second IEA international mathematics study: Purposes and design. *Journal of Curriculum Studies, 11*(3), 203–210.

Travers, K. J. (Ed.). (1985). U.S. summary report: *Second international mathematics study.* Champaign, IL: Stipes.

Wolfe, Richard G. (1986). *The IEA second international mathematics study: Overview and selected findings.* Paper presented at the annual meeting of the American Educational Research Association, San Francisco, CA. (Mimeo)

Chapter 12

The Role of Language Comprehension in Mathematics and Problem Solving

Jose P. Mestre
Department of Physics & Astronomy
University of Massachusetts

A new field of study appears to be emerging that is concerned with the influence of language on mathematical problem solving. We have chosen the term *problem processing* for this new field. For purposes of this chapter, problem processing is defined as the study of the various ways in which language can influence problem solving. As we hope to show, language proficiency interacts with problem solving in a number of different ways.

There have been several attempts to model problem-solving processes (Brown & Burton, 1978; Greeno, 1978; Newell & Simon, 1972), as well as language comprehension processes (Anderson, 1976; Kintsch & van Dijk, 1978; Schank, 1972; Winograd, 1972). However, it has only been recently that attempts have been made to integrate problem-solving and text-processing models to simulate solutions of arithmetic word problems (Briars & Larkin, 1984; Kintsch & Greeno, 1985; Riley, Greeno, & Heller, 1983). This application of text-processing models to addition and subtraction word problems built upon previous mathematics education research. This research investigated such issues as the semantic relationship among quantities in various types of word problems to model problem difficulty, the types of strategies used to solve the different types of word problems, and the types of error patterns that are commonly made by children (Carpenter & Moser, 1982; DeCorte & Vershaffel, 1981; Nesher, 1982; Nesher, Greeno, & Riley, 1982; Vergnaud, 1982).

The study of the interaction of language with problem solving takes on particular significance for students with language deficiencies, as in the case of many language minority students. In this chapter, several examples from our research are presented that illustrate how the poor comprehension skills of Hispanic bilinguals adversely affect the interpretation of both math word problems and premises similar to those used in syllogistic reasoning. Bilingual Hispanic students possessing the same level of mathematical and computational sophistication as their monolingual peers often solve word problems incorrectly. The pattern of errors suggests that language deficiencies lead to misinterpretations of word problems; the resulting solutions may be incorrect, yet mathematically consistent with the student's interpretation of the problem statement.

Although few studies have specifically investigated mathematical achievement or problem-solving proficiency with bilingual populations, there is evidence of language-related difficulties in solving mathematics problems across a number of cultures. Among the earliest studies is one conducted by the International Institute of Teachers College at Columbia University (1926). This study compared the mathematical problem solving of 12th-grade Puerto Rican bilinguals educated on the island and 12th-grade monolinguals. Despite the fact that the Puerto Rican students had been receiving mathematics instruction in English (their second language) since the fifth grade, their problem-solving ability was significantly below that of monolingual twelfth graders.

Similar studies have shown a retardation in problem-solving performance when the language of instruction was the students' weaker language (Kellaghan & Macnamara, 1968; Macnamara, 1966, 1967). Kellaghan and Macnamara's study with Irish students indicated that this result was not due to the students' inability to understand the problems since the components of the problems were separately understood by the students. According to Macnamara (1966, 1967), research evidence suggests that bilingual children keep pace with monolinguals in mechanical arithmetic, but fall behind in solving word problems.

Language proficiency appears to be a strong predictor of cognitive functioning. In a study with children from nine distinct ethnolinguistic groups, DeAvila and Duncan (1981) investigated performance on several academic, cognitive, and linguistic tasks. The results of this study showed that language proficiency was the most important predictor of achievement relative to any other factors. DeAvila and Duncan suggest the existence of a positive, monotonic relationship between linguistic proficiency and cognitive functioning, implying that academic deficiencies among the students were linguistic rather than intellectual in nature.

We present specific examples from two of our studies that investigated the ways in which language can interact with problem-solving performance among Hispanics of two age groups. First, the learning styles and problem-solving behavior of a group of Hispanic ninth graders taking beginning algebra is discussed. Among the topics covered are (a) how the students use the textbook as a learning tool, (b) ways in which the textbook is inappropriate for this group, (c) how these students solve algebra word problems, and (d) the common mathematical misconceptions exhibited by the students.

Second, we discuss the comprehension of logical premises by Hispanic college students majoring in technical fields, such as engineering. The focus of this discussion is this group's ability to comprehend complex premises similar to those that might appear in syllogistic reasoning problems, and their ability to learn and retain a simple procedure for interpreting the premises. As becomes evident, there are similarities between the strategies that students use to interpret mathematical word problems and the strategies they use to interpret complex premises. In the concluding section, we embed the issues discussed in this chapter within Cummins's (1981) theoretical framework of the role that language proficiency plays in cognitive functioning among bilinguals.

THE LEARNING OF ALGEBRA BY NINTH-GRADE HISPANICS

The concept of variable in algebra introduces a level of abstraction that goes well beyond that of arithmetic; algebra is, in fact, the "entry" course for abstract higher mathematics. Therefore, it is important to study the difficulties that students encounter in this portal course. The examples used in this section come from a study conducted with a group of ninth-grade students who were enrolled in Algebra 1. The topics that were investigated include the role of the textbook in the learning process, the selection of variable names in word problems, and the translation of word problems into equations.

Fourteen students participated in this study. These students came from two classes taught by different teachers. There were three groups of students: (a) "participating Hispanics," or 6 Hispanic students (5 male and 1 female) who were English/Spanish bilinguals and were enrolled in the mainstream curriculum, (b) "participating monolinguals," 5 English-speaking monolingual students (3 males and 2 females) from the same class, and (c) "advanced Hispanics," 3 Hispanic students (1 male and 2 female) from an advanced algebra class taught by another teacher (note that "advanced" here does not refer to language proficiency). The teacher of the first two groups (henceforth the "participating teacher") was an integral part of the study. The classroom style and textbooks used in both the Algebra 1 class and the advanced algebra class were "traditional."

The participating Hispanics were interviewed a total of eight times during the academic year, and the other two groups were each interviewed a total of four times; all interviews were audio-recorded. The students were asked to "think-aloud" while solving the computational problems or word problems chosen for that session. Following the interview session, there was a staff meeting (which included the participating teacher) during which the set of interviews was reviewed, and explanatory hypotheses for the results were proposed and discussed.

Effectiveness and Use of Textbook

A common bad habit among students enrolled in math classes is to use the textbook merely as a place to find assigned problems. This habit is lamentable in that it precludes the students from learning mathematics via written material. Yet the textbook is central to the teaching of most mathematics courses. The textbook sets the style and order in which the material is covered in the course, as well as providing students with a place to read about, and supplement, what is covered in class. It was apparent from early interviews that the majority of the students in the Algebra 1 class did not read the textbook as a means of supplemental instruction—a fact confirmed by the participating teacher. For the most part, the textbook was used by the students only as a place to find the problems assigned for exercises in class and for homework.

Despite this ineffectual use of the textbook by the students, there was a problem with the textbook that may have been largely responsible for the students' reluctance to read it. It should be noted that this "problem" with the textbook is not idiosyncratic to the particular textbook used in the Algebra 1 class; it is prevalent among the genre of mathematics textbooks used in higher mathematics. The first thing that is striking is that nearly all the existing textbooks seem to be cloned from some archetypical standard. Why this is so is not at all clear. It certainly is not a decision made by authors or publishers based upon research findings on what makes textbooks effective; there are, in fact, virtually no research findings available on the topic.

In the particular case at hand, it appears that the language and style of the textbook used in the Algebra 1 class (Dolciani & Wooton, 1973) seems more appropriate for someone with a prior knowledge of algebra than for these 14-year-old students who had no preparation beyond eighth-grade mathematics. Further, the book's attempts to draw upon "real life" situations for its word problems is of questionable pedagogic value for students from low socioeconomic levels, such as the Hispanics of our study. The following two problems from page 76 of the textbook illustrate this point:

A stock selling for $30 per share rose $2 per share each of two days and then fell $1.75 per share for each of three days. What was the selling price per share of the stock after these events?

On a revolving charge account, Mrs. Dallings purchased $27.50 worth of clothing, and $120.60 worth of furniture. She then made two monthly payments of $32.00 each. If the interest charges for the period of two months were $3.25, what did Mrs. Dallings then owe the account?

The six participating Hispanics were asked during one of the interview sessions to tell the interviewer what they thought terms such as *stock, share, revolving charge account, monthly payment,* and *interest* meant. Their responses displayed that they had little idea what these terms meant except for "interest" where several students stated that it was something that banks and stores did to make more money.

Problems like these are of questionable pedagogic value when used with students coming from low socioeconomic status because the students are more likely to be confused by the jargon, and not necessarily by the mathematics. Several students stated that even though they were not sure what some of the terms meant in the "revolving charge account" problem, they thought they could nevertheless solve the problem. Their attempts to solve this problem consisted of combining the four monetary quantities in some fashion to obtain a final, albeit incorrect answer.

A more insidious problem is related to how the students used the textbook. When students only use the textbook as a place to find assigned exercises, they may come away with the wrong interpretation for mathematical terminology. An example will clarify this point. In one of the interviews, students were asked to define various mathematical terms that had been covered recently in class. By far, the most misinterpreted term was *quotient.* Most students stated that quotient meant *answer* or *product.* Subsequent inspection of the section of the book covering quotients revealed the following instructions preceding a series of computational exercises:

Read each quotient as a product. Then state the value of the quotient.

And also:

State the value of each quotient.

It is understandable why a student reading these instructions could come away with the interpretation of *product* for the term *quotient.* The first instruction above can be taken to imply that *product* and *quotient* are interchangeable; the second instruction makes perfect sense if the word *answer* is substituted for *quotient.*

It seems that there is a lot of virgin territory in which to experiment with "plain talk" approaches as a means of (a) making textbooks more readable so that students might be more inclined to use them as learning tools, and (b) avoiding the predicament of allowing semantics to interfere with the learning of mathematics. The one virtue of this textbook, and others of its genre, is that it conveys the sense of the precision and rigor necessary to communicate in, and work with, higher level mathematics. However, there is no apparent reason why rigor cannot be introduced in a less stilted and formal style. For students with language deficiencies readability is crucial for making a textbook useful to the student.

The Processing of Word Problems

Mathematical language is fraught with definitions and conventions that help to convey the rigor and precision necessary for unambiguous communication. Math novices often have difficulties making the transition from the lack of precision inherent in natural discourse to the precision necessary in mathematical discourse. Studying the translation process from textual to symbolic representations is an excellent means of identifying difficulties caused by syntactic or semantic factors inherent in the language of mathematics. In this section, specific examples are discussed that examined the students' ability to translate mathematical statements that express a relationship between variables into mathematical equations.

Let us begin by considering the students' performance on the following four problems:

Write an expression using variables for the following statements:

1. A number added to 7 equals 18.
2. Six times a number is equal to a second number
3. Nine times a number results in 36.
4. In 7 years, John will be 18 years old.

A summary of the students' answers is shown in Table 12.1. Here, the entry "C" means that the student worked out the problem correctly; the entry "Skip" means that the interviewer skipped that problem for that student; the entry "no idea" means that the student had no idea how to solve the problem. Table 12.1 shows that students had little difficulty working out problems 1 and 3, but that problems 2 and 4 caused inordinate difficulties. Problems 1 and 3 combined generated a total of 25% errors, compared to a total of 61% errors on problems 2 and 4 combined.

There is an obvious difference between problem 4 and problems 1 and 3. In problems 1 and 3, the syntactic structure is very clear—the unknown always appears near the beginning as the noun "number." The remainder of the statement informs the student what should be done with this unknown, for example, "a number added to 7 . . . ," or "nine times a number . . . ." In contrast, the wording of problem 4 does not clearly denote the unknown; the unknown, John's current age, must in fact be deduced. Two students evidently understood problem 4, as indicated by their responses "7 + 11 = 18" and "7 + 11," but they were not able to write an equation using a variable. It appears that variations in the wording of seemingly simple problems have an observable effect on performance. In particular, if the unknown in the problem is not readily discernible, there is a higher likelihood for an error than if the unknown is clearly stated near the beginning of a problem statement.

TABLE 12.1
Student Responses

Student #	Problem Number 1	2	3	4
	Correct Response X + 7 = 18	6A = B	9Y = 36	J + 7 = 18
Participating Hispanics				
1	No Idea	6X = 12	C	Skipped
2	7 = 18	6X = 2	9 · 4 = 36	No idea
3	C	C	C	C
4	C	6N = N	C	7 + 11 = 18
5	C	6A = 2A	C	7A · 18
6	8 + 7	6N · 2N	9N + 6 = 36	8 + 9 = 18
Participating Anglos				
1	C	C	C	C
2	C	C	C	7 + A
3	C	6A = 2	C	7 + 11
4	11N + 7 = 18	6N = 2N	9A · 4N = 36	7 = 18
5	C	C	C	X − 7 = 18
Advanced Hispanics				
1	C	C	C	C
2	C	6X = X	C	Y + 7Y = 18
3	C	C	C	C

Because the wording of problem 2 and problems 1 and 3 are similar, there must be an alternative explanation for the difficulties experienced by the students in problem 2. The errors made in problem 2 are very suggestive of semantic difficulties. There appeared to be three distinct error types committed in problem 2. One error type consisted of variations of the answer "$6N = N$." The participating Hispanic who wrote this answer explained that due to the phrase "is equal" in "six times a number is equal to a second number," both numbers must be the same, and therefore N was used to represent both. We conjecture (without evidence) that this was the reasoning that caused one advanced Hispanic to write the equivalent answer, "$6X = X$." The second error type consisted of answers of the form "$6X = 2$." The participating Hispanic who wrote this answer explained that the "2" in this equation represented "the second number"; the "second" was apparently interpreted as "2." Again, we surmise (without evidence) that the participating monolingual who wrote "$6A = 2$" used similar reasoning. The third error type appeared to be a mixture of the first two error types. For example, the participating Hispanic who wrote "$6X = 12$" explained that the "12" came from multiplying the "second number" by 6. It is reasonable to assume that the three students who wrote variations of the form "$6N = 2N$" were representing "six times a number" by $6N$, and the "second number" by $2N$.

The answers to several problems indicated a proclivity toward treating variables as labels. The confusion many students possess about the distinction between variables and labels has recently attracted considerable attention. For example, consider the following problem:

Write an equation using the variables S and P to represent the following statement: "There are 6 times as many students as professors at this university." Use S for the number of students and P for the number of professors.

In this problem, approximately 35% of a large sample of nonminority engineering undergraduates committed the "variable-reversal error," which consisted of the answer "$6S = P$" (Clement, Lochhead, & Monk, 1981). Using a population of Hispanic engineering students, the frequency of the variable-reversal error was 54% (Mestre, Gerace, & Lochhead, 1982). In clinical interviews it was discovered that the variable-reversal error derived from treating S and P as labels for "students" and "professors," instead of treating them as variables to represent the *number* of students and the *number* of professors. It should be pointed out that students interviewed by Clement (1982) displayed that they were aware that there were more students than professors in the problem statement. For these students, the meaning of $6S = P$ was "six students for every one professor." The mechanism leading students to write the variable-reversed equations appeared to stem from using a sequential left-to-right translation of the problem statement. That is, "six times as many students" becomes $6S$, and because this is equal to the number of professors, $6S$ is equated to P.

In our ninth-grade algebra study, several word problems were constructed to test for the variable/label misconception. Let us first consider the three problems shown below:

   5. Mr. Smith noted the number of cars, C, and the number of trucks, T, in a parking lot and wrote the following equation to represent the situation: $8C = T$. Are there more cars or trucks in this parking lot? Why?

Write an expression with variables for the following statements:

   6. Six times the length of a stick is 24 ft.
   7. If a certain chain were four times as long it would be 36 ft.

The responses to these problems indicate a strong proclivity to treat variables as labels. In problem 5, 11 of the 14 students said there would be more cars in the parking lot as represented by the equation "$8C = T$" because of the factor of 8 in front of C. Of the remaining 3 students, one from the advanced Hispanic group correctly said that there would be more trucks. Her explanation displayed that she was using C and T correctly as variables for the number of cars and trucks. The remaining two students, both from the participating Hispanic group, did not understand the situation; one said that you could not tell whether there were more cars or trucks because the values of C and T were not given; the other said there would be an equal number of cars and trucks because of the equal sign in the equation.

In contrast to problem 5, all 14 students obtained the correct answers in both problems 6 and 7. Because both problems are about a quantity involving *length*, one might expect that students would be inclined to select the variable name L. In problem 6, 8 of these students wrote a correct equation using the letter L for the variable, and the other 6 wrote a correct equation using some other letter for the variable, such as A, or X; none of these 6, however, used the letter S for the variable, which is not unreasonable to expect since the problem also involves a stick. In contrast, the most popular variable name used in problem 7 was C, representing the word *chain*, and not L; 7 students used C, 2 used L, and the remaining used some other variable name. It is evident that the syntax of these two problems triggered (more often than not) specific variable names. That is, the manner in which problem 6 started, "six times the length" makes it clear to the student that this is a problem about length, thereby triggering the use of the letter L for the variable. However, in problem 7, the first few words, "if a chain" make it clear that this is a problem about a chain, thereby triggering the letter C to be used as the variable. Even though problem 7 is asking for a length just as problem 6 is, the student is distracted from this fact by having the references to length via the words *long* and *feet* appearing much later in problem 7. Again, it appears that the syntactic structure of these problems is largely responsible for the triggering mechanism by which a variable name is chosen, with the first important noun in the problem statement serving as the trigger. This makes it somewhat of a random process whether the student will choose a variable name to mean a label for that noun, or a quantity to be represented by the variable name.

The following four problems were constructed to investigate the students' proclivity toward committing the variable-reversal error:

8. The number of nickels in my pocket is three times more than the number of dimes.

9. The number of math books on the book shelf is equal to eight times the number of science books.

10. There are four times as many English teachers as there are math teachers at this school.

11. Last year, there were six times as many men cheating on their income tax as there were women.

TABLE 12.2
Student Responses

Student #	8	9	10	11
		Problem Number		
		Correct Response		
	$N = 3D$	$M = 8S$	$E = 4M$	$M = 6W$
		Participating Hispanics		
1	No Idea	Skipped	No Idea	No Idea
2	No Idea**	$8 \cdot 6 = 48$*	$4 \cdot 4 = 16$	$2 \cdot 6 = 12$
3	$N \cdot D =$	$8 \times S = 18$	$4 \times S = -4$	$6 \cdot W =$
4	C	C	Reversal	Reversal*
5	$3N + D$	Reversal	Reversal	Reversal
6	$3d \cdot 4 + N =$	$64M \cdot 8 \cdot 8 =$	$4MT = 20ET$	Reversal
		Participating Anglos		
1	$3N + D = X$*	$M \cdot 8S = B$*	$4 \cdot E \cdot M = T$*	Reversal*
2	a $3 \cdot B$	Reversal*	Reversal*	Reversal*
3	$N^3 \cdot D$	$b = S^8$	$E^4 \cdot M$	Reversal
4	$3N\ 10$	$N = 8$	Reversal	Reversal
5	C	C	Reversal*	Reversal*
		Advanced Hispanics		
1	C	C	C	C
2	C	C	Reversal	Reversal
3	X^3 more than N	X^8 more than N	$X^4 \cdot X = 1$*	X^6 more than N

Note: Students appear in the same order as in Table 12.1

*Upon prompting, student displayed a proper understanding of the relative sizes among the two quantities in the problem statement.

**Upon prompting, student displayed an improper understanding of the relative sizes among the two quantities in the problem statement.

The students' responses to these problems are shown in Table 12.2. The entry *Reversal* means that the student committed the variable-reversal error.

From looking at problems 8 and 9, it is clear that the syntax is such that a sequential left-to-right translation should yield the appropriate answer. For example,

"the number of nickels in my pocket"	"is (equal to)"	"three times more than the number of dimes"
N	$=$	$3D$

We therefore expect more correct answers on problems 8 and 9 than on 10 and 11, and further, we expect that those who translate problems using the sequential left-to-right method would likely get 8 and 9 correct, but 10 and 11 variable-reversed. This was borne out by the results; four students obtained both problems 8 and 9 correct as shown on Table 12.2, but of these four students, only one advanced Hispanic was able to get problems 10 and 11 correct.

There is one surprising result in Table 12.2. If we group all variable-reversal errors together with all correct responses for a particular problem, then this corresponds to all of those students who at least understood that the problem was asking for an equation relating two quantities in some specified proportion. There were a total of 4 and 6 students in this category in problems 8 and 9, respectively. However, in problems 10 and 11 there were 6 and 10 students in this category, respectively. Thus, if this aggregate correct-answer/reversal-error category is taken as a measure of "understanding" what the problem is asking, it appears that questions 10 and 11 were easier to understand than questions 8 and 9. We find this surprising since questions 8 and 9, albeit somewhat harder, have the same structure as questions 1, 2, and 3 discussed earlier in which students had little difficulty. A possible explanation for this result is that the phrasing in problems 8 and 9 is that used in formal mathematics and therefore more arcane than the colloquial phrasing in problems 10 and 11.

Finally, we would like to offer a possible explanation for the "unexpected" answers in problem 8. By unexpected answers we mean the three answers with the inequality "greater than," and the four answers that were not in the form of an equation, e.g., "$N^3 \times D$" and "$N \times D = .$" It should first be noted that these types of answers did not occur with anywhere near the same frequency in problem 9—a problem fairly equivalent in structure to problem 8. It appears that the biggest difference between these two problems is that in problem 8, the word *equal* was never explicitly used as was the case in problem 9; the equality in problem 8 had to be deduced from context. Those students who wrote their answers as inequalities seem to have interpreted the phrase "is three times more than" as a statement of inequality rather than a statement of equality.

The examples provided in this section illustrate that a word problem's syntax can affect comprehension and performance. There is also evidence that students often rely on natural language to interpret mathematical terminology, such as representing a variable referred as "second number" in a problem with "$2N$" or the value "2." Finally a very common practice exhibited by students in translating word problems into equations is to process the word problem using a sequential left-to-right method with little regard to whether the equation thus generated matches the mathematical meaning of the problem.

THE ROLE OF LANGUAGE IN PREMISE COMPREHENSION

In this section, the results of a study that investigated the comprehension of complex premises containing negations by both Hispanic and monolingual undergraduates is summarized. As becomes apparent, college students attempting to interpret complex premises exhibit problem-solving behavior that is quite similar to the problem-solving behavior exhibited by the ninth-grade subjects discussed in the previous section.

Errors Patterns in Comprehending Premises

The types of premises that were chosen contained different number and types of negations. There were two reasons for selecting premises containing negations. First, a considerable number of research studies indicate that sentences containing negations are harder to comprehend than affirmatively phrased sentences (Just & Carpenter, 1971; Mehler, 1963; Slobin, 1965; Trabasso, Rollin, & Shaughnessy, 1971; Wason, 1959, 1961). Research findings also indicate that increasing the number of negations results in successive decrements in comprehension (Johnson-Laird, 1970; Legrenzi, 1970; Sherman, 1973, 1976) with sentences containing three or more negations beyond the normal comprehension ability of subjects. Given the difficulties that subjects have interpreting sentences containing negations, a comprehension task using premises containing different number and types of negations appeared promising for identifying some of the common difficulties and misconceptions that interfered with gleaning the appropriate meaning from negated statements.

The second reason for our interest in negations derives from a lack of parallelism between English and Spanish in the meaning of certain doubly negated constructions. The English phrasings of certain Spanish double-negative statements contain only a single negative. In Spanish, these doubly negated constructions retain a negative meaning instead of reverting to an affirmative meaning, as would be the case in grammatically correct English. For example, the Spanish translation of "I do not want anything" is *yo no quiero nada*. Translated literally into English, this Spanish statement would become "I do not want nothing." The Spanish negations *no* and *nada* (meaning no and nothing) result in an overall negative meaning when they appear together in a sentence. This lack of parallelism between English and Spanish gives rise to the question, Do monolinguals and Hispanics exhibit a different pattern of performance when interpreting premises containing double negations? Therefore, one of the goals of the study was to attempt to answer this question.

The study (Mestre, Hardiman, Gerace, & Well, in press) was conducted with both monolinqual and Hispanic undergraduates majoring in technical fields, such as engineering. The task that these subjects performed consisted of reading a premise that contained either one, two, or three negations and then selecting from a set of four affirmatively phrased statements the one choice that conveyed the equivalent meaning as the premise. Eight different types of premises were used:

1. Not all clerks working at the Fitzgerald Company are male
2. Not all clerks working at the Fitzgerald Company are not male
3. It's not true that all clerks working at the Fitzgerald Company are male
4. It's not true that all clerks working at the Fitzgerald Company are not male
5. It's not true that not all clerks working at the Fitzgerald Company are male
6. It's not true that not all clerks working at the Fitzgerald Company are not male
7. It's not true that some clerks working at the Fitzgerald Company are male
8. It's not true that some clerks working at the Fitzgerald Company are not male

The following set of affirmatively phrased, multiple-choice answers were provided with all of these eight premises:

1. All clerks working at the Fitzgerald Company are male
2. Some clerks working at the Fitzgerald Company are male
3. All clerks working at the Fitzgerald Company are female
4. Some clerks working at the Fitzgerald Company are female

Although no significant differences between monolinguals and Hispanics were found in comprehending the premises containing double negations, there were a number of interesting findings. First, overall performance on the eight premise types averaged over all subjects was approximately 50%, with Hispanics scoring below monolinguals; this level of performance is rather poor in view of the expected score of 25% if one simply randomly guessed the answer.

Second, there was a related series of erroneous strategies that students commonly used to paraphrase the premises into affirmative statements. Over 85% of all errors consisted of the application of a simple fallacious strategy. This strategy consisted of processing the negations from left to right, using rules governing the interpretation of natural discourse to paraphrase the negative portions as they occur. For example, the premises "Not all clerks . . . are male" and "It's not true that all clerks . . . are male" were paraphrased into "Some clerks . . . are male." These interpretations would be perfectly reasonable in the context of natural discourse. The addition of a negation at the end of a premise introduced a twist in the procedure. Both "It's not true that all clerks . . . are not male" and "Not all clerks . . . are not male" were interpreted to mean "Some clerks are female." This was accomplished by first converting "not all" to "some" and then changing "not male" to "female."

These erroneous patterns were quite consistent across subjects. It is also clear why the strategy described above was so common. It appears natural to process premises sequentially from left to right as one reads them, as was the case when translating word problems into equations. Further, it also appears natural to use the rules that govern the comprehension of natural discourse for interpreting logical premises. For example, if the statement "not all clerks working at such-and-such company are male" were used in a conversation, the intended meaning would most probably be that "some clerks are male," versus the strictly logical meaning, "some clerks are female."

Effect of an Intervention Strategy on Performance

Because these misconceptions were so prevalent, we were interested in ascertaining whether they could be overcome with a short intervention strategy. The intervention strategy was implemented on a videotape lasting approximately one half hour. In designing the intervention strategy, an attempt was made to take into account research findings from cognitive research studies in various domains. The intervention provided students with a procedure for parsing the premises using a right-to-left strategy. An attempt was also made to directly address the misconceptions discussed above (e.g., replacing "not all" with "some") to see if they could be supplanted with the appropriate understanding; this focus was reinforced by research findings from domains such as physics (Champagne, Klopfer, & Gunstone, 1982), algebra (Clement, 1982; Mestre et al., 1982) and statistics (Pollatsek, Lima, & Well, 1981; Tversky & Kahneman, 1977), which reveal that misconceptions are deeply lodged. Finally, the subjects had an opportunity to prectice the right-to-left parsing strategy during the course of the videotaped lesson, since there is evidence that subjects learn a procedure best by precticing it (Anzai & Simon, 1979).

The real measure of the effectiveness of an intervention strategy is the pervasiveness of its effects over time. An intervention strategy would be of limited pedagogical value if performance could be improved immediately following its implementation, only to have this improvement disappear after a period of time. To test the long-term effects of the intervention strategy, the subjects were tested three different times: immediately following the videotaped lesson, 1 week following, and 6 months following.

The effect of the intervention strategy on performance was quite revealing. Of the 20 monolingual subjects participating in the study, all achieved ceiling-level performance immediately following the intervention strategy on a 32-item assessment made up of four sets of eight premises similar to those shown earlier. One week later, all 20 subjects had retained ceiling-level performance. Six months following the intervention strategy, 14 of the monolingual subjects returned to take the assessment one last time; all but one displayed ceiling-level performance.

For the Hispanic subjects, the intervention strategy proved to be less effective. Of the 17 Hispanic subjects who took the assessment instrument immediately following the videotaped lesson, 11 reached ceiling-level performance and 6 scored below 70%. These 11 Hispanics maintained ceiling-level performance in the assessment administered 1 week following the lesson. However, 6 months following the lesson, only 3 of the 8 ceiling-level Hispanics who returned to take the assessment performed at ceiling-level; the other 5 scored below 80%.

The question that immediately comes to mind is, Why was the intervention strategy so effective for the monolingual subjects, but not as effective for the Hispanic subjects? A possible answer to this question can be found in the analysis of the error patterns of the Hispanic subjects following the intervention strategy. Over 85% of the errors committed by the six Hispanic subjects who performed poorly immediately following the lesson resulted from continued reliance on a left-to-right processing strategy and using rules governing natural discourse comprehension to interpret negated phrases (e.g., simply replacing "not all" with "some"). The error patterns for those Hispanics who performed at ceiling-level immediately following the intervention strategy, but who fell below 80% after 6 months, were almost identical to those of the Hispanics who performed poorly immediately following the lesson. Despite the fact that many of the Hispanic subjects successfully abandoned the original fallacious strategies after the lesson, a substantial number of them subsequently reverted to these erroneous strategies after a period of time had elapsed following the lesson.

This analysis of the error patterns explains what happened, but not why it happened. One explanation for reverting to fallacious strategies could be related to the persistence of misconceptions. However, since the monolingual subjects were able to overcome their misconceptions, there must be another explanation. The explanation that makes the most sense is that the Hispanic group was less prepared in English proficiency than the monolingual group. For example, the average SAT-Verbal score for the monolingual group was 502, compared to only 333 for the Hispanic group. This relative language deficiency might prove a handicap in an intervention strategy that was delivered verbally via a videotaped lesson and relied on subjects' ability to distinguish subtleties in meaning among some rather complex premises.

LANGUAGE AS A MEDIATOR OF PROBLEM SOLVING

The preceding examples suggest that language proficiency mediates cognitive functioning. One can categorize four different forms of "language proficiencies" that can influence problem solving in a technical domain like mathematics: Proficiency with language in general, proficiency in the technical language of the domain, proficiency with the syntax and usage of language in the domain, and proficiency with the symbolic language of the domain. The first of these, general language proficiency, can be assessed by typical language proficiency instruments, such as the verbal portion of the SAT. Since problem solving involves reading and comprehending the problem under consideration, ability to understand written text is of paramount importance. A related adverse effect that general language deficiencies have on problem solving is the speed with which language-deficient students solve problems; the disparity in performance on *timed* word problem-solving tasks between Hispanics and monolinguals is often due to the fact that Hispanics read at much slower speeds and reach fewer problems (Mestre, 1986)

Second, proficiency in the technical language of the domain can have an influence on problem-solving performance. The vocabulary and syntax used in a technical domain like mathematics have equivalent counterparts in natural language; however, the meaning of words or phrases can be drastically different depending on whether we are within a mathematical context, or a natural language context. For example, the word *product* has both a mathematical and a natural discourse meaning; in natural discourse a *product* is an item sold in a store, whereas in mathematics, *product* is the result of the operation of multiplication.

The third way in which language proficiency can influence problem solving concerns the way that mathematical and natural language are structured to form text: the problem solver must be able to distinguish when a word is being used mathematically and when it is not. In addition, the solver must understand how the parts of the problem are related to each other in a mathematical sense. For example, when attempting to solve the "students and professors" problem cited earlier, many students use a strict left-to-right parsing strategy. This strategy does not work in translating the problem from text to symbolic language because the mathematcal meaning of the problem cannot be related to the problem's syntax in this fashion. Clearly, relying on the surface features of a text can lead to misinterpretations.

The fourth language factor that influences problem solving in a technical domain is proficiency within the symbolic language of the domain. A domain like mathematics has a very rich and efficient symbolic language with its own grammar. For example, the mathematical statement "$2<X<8$" is totally "grammatical" within the symbolic language of mathematics; however, the variation, "$2<X>8$" is ambiguous since it states that $X>2$ *and* $X>8$, both of which cannot always be satisfied simultaneously. Further, the variation "$2>X>8$" is not grammatical since it implies that $X<2$ *and* $X>8$, either of which contradicts the other. Thus, proficiency within the symbolic language of the domain will influence problem-solving performance.

Two of these language proficiency categories, namely technical language proficiency and symbolic language proficiency, are domain dependent. To develop problem-solving skills in a specific domain, one must become proficient in these two "languages" within that domain. However, there is a catch to learning these two languages: proficiency in the first category, namely English language proficiency, is likely to mediate the learning of the technical and symbolic languages. This means that individuals with below-average English language proficiency are at a relative disadvantage. Because the language proficiency level of bilinguals is often below that of monolinguals, bilingual students face a hurdle in developing problem solving and in learing from written or oral presentations.

Cummins (1981) posits an interesting hypothesis that relates the level of language proficiency and cognitive functioning for bilingual students. Cummins's hypothesis states that ''there may be a threshold level of linguistic competence which bilingual children must attain both in order to avoid cognitive deficits and to allow the potential beneficial aspects of becoming bilingual to influence their cognitive growth'' (p. 229). Cummins does not define the threshold level in absolute terms since it is likely to vary depending on the bilingual's stage of cognitive development, and on the academic demands of the different stages of schooling.

Cummins does define three types of bilingualism. The first, *semilingualism,* is characterized by a below-threshold level of linguistic competence in *both* languages. In semilingualism, both languages are sufficiently weak to impair the quality of interaction the student can have with the educational environment. The negative aspects of semilingualism are no longer present in *dominant bilingualism,* characterized by an above-threshold level of competence in *one* of the two languages. Dominant bilingualism is supposed to have neither a positive nor a negative effect on cognitive development. The last category, *additive bilingualism,* is one that has positive cognitive effects. Additive bilingualism is characterized by above-threshold competence in both languages.

Many of the language minority students at all levels of education in the United States are semilingual. Semilingual students are often raised in the mainland, speaking their first language at home and in their neighborhoods, and only receive English instruction at school. Most mandated bilingual programs in primary and secondary schools are ''transitional'' programs. This means that schools have 3 years to teach enough English to these limited-English proficiency students so they can be mainstreamed. The result of most of these programs is students who are linguistically competent to function in the mainstream curriculum, but who are not linguistically proficient at a level where they can favorably compete with monolingual students. Since little to no formal training is given to these students in Spanish, and since their English skills receive little support outside of the school setting, it is not surprising that bilingual programs often turn out semilinguals.

Similar to Cummins's linguistic threshold, Burns, Gerace, Mestre, and Robinson (1983) have proposed a ''technical threshold'' that students must cross before becoming proficient at problem solving in a technical domain. The technical threshold consists of reaching a certain proficiency level in what we have referred to above as technical language and symbolic language. As we argued earlier, bilingual students face the additional burden of becoming highly proficient in English before they can successfully reach technical threshold.

In summary, language proficiency affects problem solving, especially in tasks that require substantial amounts of language processing, as in solving math word problems. Bilingual students whose language proficiency level is below that of their monolingual counterparts face disadvantages, such as misinterpreting the meaning of problems, and reading speeds slow enough to interfere with performance in timed problem-solving tasks. It is important that educators become aware of the role that language plays in problem solving, and equally important that researchers continue to study the role of language in problem solving in order to make the necessary pedagogical improvements.

ACKNOWLEDGMENTS

The work reported herein was supported by National Institute of Education Grant #G-83-0072 and Contract #400-81-0027. The contents of the chapter do not necessarily reflect the position, policy, or endorsement of the National Institute of Education. Many thanks to Drs. W.J. Gerace and P.T. Hardiman for their suggestions.

REFERENCES

Anderson, J. R. (1976). *Language, memory, and thought*. Hillsdale, NJ: Lawrence Erlbaum Associates.

Anzai, Y., & Simon, H. A. (1979). The theory of learning by doing. *Psychological Review, 86,* 124–228.

Burns, M., Gerace, W., Mestre, J. P., & Robinson, H. (1983). The current status of Hispanic technical professionals: How can we improve recruitment and retention. *Integrateducation, 20,* 49–55.

Briars, D. J., & Larkin, J. H. (1984). An integrated model of skill in solving elementary word problems. *Cognition & Instruction, 1,* 245–296.

Brown, J. S., & Burton, R. R. (1978). Diagnostic models for procedural bugs in basic mathematical skills. *Cognitive Science, 2,* 155–192.

Carpenter, T. P., & Moser, J. M. (1982). The development of addition and subtraction problem-solving skills. In T. P. Carpenter, J. M. Moser, & T. Romberg (Eds.), *Addition and subtraction: A cognitive perspective* (pp. 9–24). Hillsdale, NJ: Lawrence Erlbaum Associates.

Champagne, A. B., Klopfer, L. E., & Gunstone, R. F. (1982). Cognitive research and the design of science instruction. *Educational Psychologist, 17,* 31–53.

Clement, J. (1982). Algebra word problem solutions: Thought processes underlying a common misconception. *Journal for Research in Mathematics Education, 13,* 16–30.

Clement, J., Lochhead, J., & Monk, G. S. (1981). Translation difficulties in learning mathematics. *American Mathematical Monthly, 88,* 286–290.

Cummins, J. (1981). Linguistic interdependence and the educational development of bilingual children. *Review of Educational Research, 49,* 222–251.

DeAvila, E. A., & Duncan, S. F. (1981). The language minority child: A psychological, linguistic, and social analysis. In J. E. Alatis (Ed.), *Georgetown University Roundtable on Languages and Linguistics, 1980: Current Issues in Bilingual Education* (pp. 104–137). Washington DC: Georgetown University Press.

DeCorte, E., & Vershaffel, L. (1981). Children's solution processes in elementary arithmetic problems: Analysis and improvement. *Journal of Educational Psychology, 6,* 765–779.

Dolciani, M. P, & Wooton, W. (1973). *Modern algebra: Structure and method. Book I*. Boston, MA: Houghton Mifflin.

Greeno, J. G. (1978). Understanding and procedural knowledge in mathematics education. *Educational Psychologist, 12,* 268–283.

International Institute of Teachers College. (1926). *A survey of the public educational system of Puerto Rico*. New York: Bureau of Publications, Teachers College, Columbia University.

Johnson-Laird, P. N. (1970). Linguistic complexity and insight into a deductive problem. In G. B. Flores d'Arcais & W. J. M. Levelt (Eds.), *Advances in psycholinguistics* (pp. 334–343). Amsterdam: North-Holland.

Just M. A., & Carpenter, P. A. (1971). Comprehension of negation with quantification. *Journal of Verbal Learning & Verbal Behavior, 10,* 244–253.

Kellaghan, J., & Macnamara, J. (1967). Reading in a second language in Ireland. In M. D. Jenkinson (Ed.), *Reading instruction, an international forum; proceedings* (pp. 231–240). Newark, DE: International Reading Association.

Kintsch, W., & van Dijk, T. A. (1978). Toward a model of test comprehension and production. *Psychological Review, 85,* 363–394.

Kintsch, W., & Greeno, J. G. (1985). Understanding and solving word arithmetic problems. *Psychological Review, 92,* 109–129.

Legrenzi, P. (1970). Relations between language and reasoning about deductive rules. In. G. B. Flores d'Arcais & W. J. M. Levelt (Eds.), *Advances in psycholinguistics* (pp. 322–333). Amsterdam: North-Holland.

Macnamara, J. (1966). *Bilingualism in primary education*. Edinburgh: Edinburgh University Press.

Macnamara, J. (1967). The effects of instruction in a weaker language. *Journal of Social Issues, 23,* 121–135.

Mehler, J. (1963). Some effects of grammatical transformations on the recall of English sentences. *Journal of Verbal Learning & Verbal Behavior, 2,* 346–351.

Mestre, J. P. (1986). Teaching problem-solving strategies to bilingual students: What do research results tell us? *International Journal of Mathematical Education in Science and Technology, 17,* 393–401.

Mestre, J. P., Gerace, W. J., & Lochhead, J. (1982). The interdependence of language and translational math skills among bilingual Hispanic engineering students. *Journal of Research in Science Teaching, 19,* 339–410.

Mestre, J. P., Hardiman, P. T., Gerace, W., & Well, A. (in press). Comprehending premises: Effects of negations and training among Anglos and Hispanics. *Journal of the National Association for Bilingual Education* (NABE Journal).

Nesher, P. (1982). Levels of description in the analysis of addition and subtraction. In T. P. Carpenter, J. M. Moser, & T. Romberg (Eds.), *Addition and subtraction: A cognitive perspective* (pp. 25–38). Hillsdale, NJ: Lawrence Erlbaum Associates.

Nesher, P., Greeno, J. G., & Riley, M. S. (1982). The development of semantic categories for addition and subtraction. *Experimental Studies in Mathematics, 13,* 373–394.

Newell, A., & Simon, H. A. (1972). *Human problem solving*. Englewood Cliffs, NJ: Prentice-Hall.

Pollatsek, A., Lima, S. D., & Well, A. (1981). Concept of computation: Students' understanding of the mean. *Educational Studies in Mathematics, 12,* 191–204.

Riley, M. S., Greeno, J. G., & Heller, J. I. (1983). Development of children's problem-solving ability in arithmetic. In H. P. Ginsburg (Ed.), *The development of mathematical thinking* (pp. 152–196). New York: Academic Press.

Schank, R. C. (1972). Conceptual dependency: A theory of natural language understanding. *Cognitive Psychology, 3,* 552–631.

Sherman, M. A. (1973). Bound to be easier? The negative prefix and sentence comprehension. *Journal of Verbal Learning and Verbal Behavior, 12,* 76–84.

Sherman, M. A. (1976). Adjectival negation and the comprehension of multiply-negated sentences. *Journal of Verbal Learning & Verbal Behavior, 15*, 143–157.

Slobin, D. I, (1965). Grammatical transformations and speed of understanding. *Journal of Verbal Learning & Verbal Behavior, 5*, 219–227.

Trabasso, T., Rollin, H., & Shaughnessey, E. (1971). Storage and verification shapes in processing concepts. *Cognitive Psychology, 2*, 231–289.

Tversky, A., & Kahneman, D. (1977). Judgment under uncertainty: Heuristics and biases. In P. N. Johnson-Laird & P. C. Wason (Eds.), *Thinking: Readings in cognitive science* (pp. 326–337). Cambridge: Cambridge University Press.

Vergnaud, G. (1982). A classification of cognitive tasks and operations of thought involved in addition and subtraction problems. In T. P. Carpenter, J. M. Moser, & T. Romberg (Eds.), *Addition and subtraction: A cognitive perspective* (pp. 39–59). Hillsdale, NJ: Lawrence Erlbaum Associates.

Wason, P. C. (1959). The processing of positive and negative information. *Quantitative Journal of Experimental Psychology, 11*, 92–107.

Wason, P. C. (1961). Response to affirmative and negative binary statements. *Journal of Psychology, 52*, 133–142.

Winograd, T. (1972). *Understanding natural language.* New York: Academic Press.

Chapter 13

Linguistic Features
of Mathematical Problem Solving:
Insights and Applications

George Spanos, Nancy C. Rhodes, Theresa Corasaniti Dale,
and JoAnn Crandall
Center for Applied Linguistics
Washington, DC

INTRODUCTION

Language Proficiency and Mathematics Achievement

Students whose home language is not English must acquire a substantial level of English proficiency to be able to participate effectively in U.S. schools. Cummins (1981) postulates that there exists a minimal level of linguistic competence—a threshold—that a student must attain to function effectively in cognitively demanding, academic tasks. This threshold of cognitive academic language proficiency (CALP) can take between 5 and 7 years to develop in a student's second language.

Burns, Gerace, Mestre, and Robinson (1983), expanding this notion, have proposed a technical threshold to explain the strong positive correlation they found between the English language proficiency of Hispanic college students enrolled in technical classes and their performance on mathematics tests. They argue that students who lack language skills or the cognitive skills to solve technical problems, or both, have not yet reached the level required to comfortably read and solve technical problems. They believe that materials and teaching methods need to be developed that will assist students in attaining this threshold so that they can participate effectively in the cognitively demanding tasks required in mathematics and science classes.

Dawe (1984) suggests an interesting perspective on the role that language plays in the performance of cognitively demanding tasks in mathematics. He entertains the hypothesis that students must reach a threshold level of proficiency in cognitive academic mathematics proficiency (CAMP). CAMP consists of cognitive knowledge (mathematical concepts and how they are applied) embedded in a language specifically structured to express that knowledge. The threshold level for CAMP, which students must achieve to perform math tasks, consists of proficiency both in mathematics and in math language. This tentative proposal lends support to Mestre's (1986) suggestion that math instruction for limited-English-proficient students (if not for all students) should follow an approach that integrates rather than separates math skills and language skills.

A growing body of research suggests a close relationship between language proficiency and mathematics achievement. Studies with monolingual English speakers have revealed a high positive correlation between mathematics achievement and English reading ability (Aiken, 1971). Duran (1979) found a similar positive correlation between the reading comprehension skills of Puerto Rican college students and their performance on deductive reasoning problems in English and Spanish, with a similar pattern across both languages. Mestre (1981) found a strong correlation between language proficiency and math performance among college Hispanic engineering students as did Cossio (1978). Cuevas (1984) has shown that language is a factor both in the learning of mathematics and in the assessment of mathematics achievement. Further, several researchers have found that language minority students frequently do not understand the language used to present math test problems (Crandall, Dale, Rhodes, & Spanos, in press; DeAvila & Havassy 1974; Moreno 1970; Ramirez & Gonzalez, 1972).

Thorndike (1912) said, ''Our measurement of arithmetic is a measure of two things: sheer mathematical knowledge on the one hand, and acquaintance with language on the other.'' Language skills are the vehicles through which students learn, apply, and are tested on math concepts and skills. Unfortunately, the language of mathematics is often too difficult for many students. Consider the following elementary algebra problem:

> Find a number such that seven less than the number is equal to
> twice the number minus 23.

Although the solution to the problem is straightforward once the proper equation for the solution has been derived, the linguistic skills required to reach that point are rather sophisticated. First, one must understand that *such that* relates *seven less than the number* to *twice the number minus 23*. It is also necessary to understand that *a number* and the subsequent *the number* (which is repeated twice) refer to the same number. Finally, the phrase *seven less than the number* is syntactically confusing. It often leads the student to write $7 - n$, when the reverse, $n - 7$, is required. The problem also assumes that the student understands that the phrase *is equal to* signals an equation. Gerace and Mestre (1983) have demonstrated that these types of constructions cause difficulties for ninth-grade algebra students with limited English proficiency. Crandall et al. (in press) have found similar problems among college developmental algebra students.

Halliday (1975) has suggested that mathematics comprises a unique linguistic register with special features that must be mastered by students of mathematics and mathematics-related disciplines. He defines a linguistic register as "a variety of language that is oriented to a particular context, to a certain type of activity, involving certain groups of people, with a certain type of activity, involving certain groups of people, with a certain rhetorical force" (p. 5). Thus, the mathematics register would be the variety of language oriented to mathematics activities and would comprise the various linguistic forms, their meanings and uses, that appear in the context of these activities.

Despite the evidence and conjectures cited above, limited-English-proficient students are traditionally "mainstreamed" into mathematics classes before they are placed into other academic content classes on the invalid assumption that math is "language independent." To the contrary, students would benefit from an approach that simultaneously taught language and mathematics content.

An Applied Research Project

This close and constant relation between language proficiency and mathematics achievement has motivated an applied research project to (a) investigate linguistic features that pose difficulties in understanding and solving algebra problems and (b) develop materials that would address these linguistic difficulties while increasing students' understanding of basic algebra concepts. Working collaboratively with mathematics departments at three colleges (Miami-Dade Community College, Northern Virginia Community College, and Metropolitan State College), we have developed an approach and created materials to assist students in becoming proficient in using the language of math (Crandall, Dale, Rhodes, & Spanos, 1987). In first year algebra, students review basic math and begin working with increasingly abstract and symbolic language. For that reason, and because algebra is a gatekeeping course for all scientific and technical disciplines, attention was focused on developmental algebra.

Data collection procedures at the three colleges included analyzing texts and tests, and most important, recording student discussions as they worked together to solve math problems. Forty-six developmental and college algebra students, some Hispanic, some limited-English-proficient, and some native English-speaking, were interviewed, usually in groups of two or three students. We asked them to complete general information questionnaires and then to talk together as they attempted to solve a series of algebra problems. These think-aloud sessions, which lasted about an hour each, were audio-taped and later transcribed. The data from these problem-solving sessions have been a particularly important source of information on the difficulties students face in working through arithmetic and algebra word problems and on the particular linguistic features that are most problematic.

When students were first asked, individually, to explain what they were thinking about and how they were approaching standard algebra problems, they were often unable to do so, since much of their difficulty with math resided in their inability to use math language or to verbalize their thinking and problem-solving strategies. However, when placed in groups of two or three students and asked to cooperatively solve a problem, such verbalization came much more naturally. Students suggested ways in which the problem might be understood or stated in mathematical terms for solution or they asked questions of other students.

A LINGUISTIC MODEL FOR CATEGORIZING FEATURES OF THE MATHEMATICS REGISTER

As a result of the interviews, text and test analyses, and problem-solving sessions, we were able to identify a number of types of linguistic difficulties students face in dealing with the mathematics register. In this section, we cite specific syntactic, semantic, and pragmatic features from our research that caused difficulties for beginning algebra students. In the next section, we discuss the broader problem-solving contexts in which such difficulties have been observed to arise.

It is useful in this regard to appeal to an important trichotomy proposed by Morris (1955) and adopted by Carnap (1955) to categorize the linguistic features that attend particular scientific domains. In his seminal work in semiotics (the systematic study of linguistic and nonlinguistic signs) Morris (1955) distinguished between the following three sub-studies:

syntactics: the study of how linguistic signs, or symbols, behave in relation to each other, e.g., the formal relationship that obtains between active and passive verb forms in English;

semantics: the study of how linguistic signs behave in relation to the objects or concepts they refer to (their denotations) or their senses (their connotations), e.g., the relationship between the English word *star* and the numerous heavenly and earthly objects to which it may refer and the several senses it may have;

pragmatics: the study of how linguistic signs are used and interpreted by speakers of natural languages in specific contexts of use, e.g., the relationship in English between the words ''I promise'' and a speaker's intentions to perform a future action or a hearer's expectations that the action will be performed.

In applying these definitions to the signs and symbols that characterize the mathematics register, we must come to grips with both the notational forms created for use by mathematicians and the English equivalents of these forms as they appear in mathematical discourse, including student problem sets. Thus, the source of many of the difficulties discussed in this section can be traced at the outset to the complex interplay between syntactic, semantic, and pragmatic features that occurs when students attempt to verbalize or interpret mathematical rules and concepts in English. Using this categorization of features, we can cite specific problematic features in the outline given in Table 13.1.

Discussion of Syntactic Features

Knight and Hargis (1977) have pointed out that comparative structures are especially useful and widespread in mathematical discourse. Unfortunately for many language minority students, the English comparative system can be rather complex. Structures such as *greater than, less than, n times as much as, as . . . as* are often confusing at the syntactic level because they require that the student master complex patterns that relate to specific meanings in a variety of ways.

Consider, for example, the fact that the following sentences are paraphrases of one another:

1. Triangle A is as large as Triangle B.
2. Triangles A and B are equal in size.
3. Triangle A and Triangle B are the same in size.

The same problem (cited by Munro, 1979) exists with the syntactic patterns required by prepositional phrases and the passive voice. Consider, for example, sentences 4–6:

4. Four divided into nine equals nine-fourths.
5. Nine divided by four equals nine-fourths.
6. If nine is divided by four, nine-fourths results.

As with the comparative structures in 1–3, sentences 4–6 are paraphrases that require a rather advanced facility with two-word verbs ending in prepositional particles and corresponding passives. The potential for confusion becomes evident when one considers that textbook writers and instructors are apt to employ a variety of patterns in their exercises and lectures that, although stylistically desirable, are beyond the competence of a sizable number of students, both native and foreign.

A further syntactic difficulty is related to the lack of a regular one-to-one

TABLE 13.1
Syntactic, Semantic and Pragmatic Features of the Mathematics Register

---

SYNTACTIC FEATURES

I. COMPARATIVES

greater than/less than	as in	all numbers greater than 4
n times as much as	as in	Hilda earns six times as much as I do. Hilda earns $40,000 a year. What do I earn?
as. . .as	as in	Wendy is as old as Jack. Jack is three years older than Frank. Frank is 25. How old is Wendy?

II. PREPOSITIONS

divided *into*   as in   four (divided) into nine (9/4 or 9 ÷ 4)
divided *by*   as in   four divided by nine   (4/9 or 4 ÷ 9)
by   as in   two is multiplied by itself three times (multiplication)

vs.

x exceeds two by seven (addition)

III. PASSIVE VOICE

as in x is defined to be greater than or equal to zero.
When 15 *is added* to a number, the result is 21. What is the number?

IV. REVERSAL ERRORS

Examples: The number a is five less than the number b.
correct equation: $a = b - 5$
incorrect equation: $a = 5 - b$ or $a - 5 = b$

There are five times as many students as professors in the mathematics department.
correct equation: $5p = s$
incorrect equation: $5s = p$

V. LOGICAL
CONNECTORS

if. . .then   as in   If a is positive then $-a$ is negative.
if and only if   as in $a + b = c$ if and only if $b + a = c$.
given that   as in   Given that $a = 0$, $a \times b = 0$.

SEMANTIC FEATURES

I. LEXICAL
A. NEW TECHNICAL VOCABULARY

additive inverse	coefficient	denominator
binomial	monomial	polynomial

B. NATURAL LANGUAGE VOCABULARY WHICH HAS A DIFFERENT MEANING IN MATHEMATICS

square	rational	irrational
power	equality	inequality

C. COMPLEX STRINGS OF WORDS OR PHRASES
least common denominator
negative exponent
the quantity, $y + 3$, squared

(*Continued*)

D. SYNONYMOUS WORDS AND PHRASES

For addition: add, plus, combine, sum,
more than, and increase by

For subtraction: subtract, minus, differ(ence),
less than, and decreased by

E. SYMBOLS AND MATHEMATICAL NOTATION AS "VOCABULARY"

$$= \qquad > \qquad \geq \qquad (\ )$$
$$\sim \qquad < \qquad \leq \qquad [\ ]$$

II. REFERENTIAL

A. ARTICLES/PRE-MODIFIERS

a number. . .the number   as in   Five times a number is two more than ten times the number.

one number. . .another number   as in   One number is ten times another number. If the first number is 7, find the second number.

B. VARIABLES

Example: There are five times as many apples as bananas in the fruit bowl.

correct equation: $5b = a$, where b refers to the *number* of bananas and a refers to the *number* of apples

III. VAGUENESS IN PROBLEMS AND DIRECTIONS

Example: Food expenses take 26% of the average family's income. A family makes $700 a month. How much is spent on food?

(That month?   In a year?   What is being asked for??)

IV. SIMILAR TERMS, DIFFERENT FUNCTIONS

less	vs.	less than
the square	vs.	the square root
divided by	vs.	divided into
multiply by	vs.	increased by

PRAGMATIC FEATURES

I. EPISTEMOLOGICAL

A. Lack of Experience or Knowledge

e.g., market-place concepts, e.g., discounts, cost, selling price, markup, wholesale, retail, sales tax rates

B. Restricted Experience or Knowledge

e.g., attempt to substitute known quantities like local tax rates for tax rates referred to in word problems

C. Conflicting Experience or Knowledge

e.g., inability to solve for tax rate because in practical experience this is a given fact

D. Contradictory Experience or Knowledge

e.g., discrepancy between the way sales tax is rounded off on actual sales tax charts and the way it is rounded off conventionally

II. TEXTUAL

A. Lack of Real Life Objects or Activities (Realia) in Math Curricula

B. Lack of Natural Interaction

correspondence between the strings of mathematical symbols in numerical expressions and equations and the strings of words used in their natural language counterparts. For example, although sentences 4–6 above are all symbolically equivalent to the equation $9/4 = 9/4$, students often attempt to duplicate the surface word order in rendering the equations into symbolic notation. Such an attempt would result in a reversal error for sentence 4, since the word order might lead one to write $4/9 = 9/4$. Similar errors noted by Crandall et al. (in press) are evident in sentence pairs 7a, 7b and 8a, 8b below:

7a. The number a is five less than the number b.
7b. $a = 5 - b$ (instead of $a = b - 5$)

8a. There are three times as many girls as boys at this university.
8b. $3g = b$ (instead of $3b = g$)

Logical connectors, which are "words or phrases which carry out the function of marking a logical relationship between two or more basic linguistic structures (and) serve a semantic, cohesive function indicating the nature of the relationship between parts of a text" (Kessler, Quinn, & Hayes, 1986, p. 14), are widely used in mathematics texts to develop and concatenate mathematical structures. Some of these connectors include *if . . . then, if and only if, because, that is, for example, such that, but, consequently, and either . . . or.*

When students read mathematics texts, they must be able to recognize logical connectors and the situations in which they appear. They must know which situation is signaled—similarity, contradiction, cause/effect or reason/result, chronological or logical sequence. On the level of syntax, they have to know where logical connectors appear in a sentence (clause initial, medial, or final) and they must be aware that some connectors can only appear in one position, but others can appear in two or all three positions. Moreover, a change in position can signal a change in meaning.

Examples abound in math texts where logical connectors are used to introduce definitions and properties—concepts that students must understand and apply to solve problems. These connectors most often appear in complex statements using both words and symbols, as in the following examples from Dolciani and Wooton (1970, p. 77):

1. If a is a positive number, then $-a$ is a negative number;
   if *a is a negative number, then* $-a$ is a positive number;
   if a is 0, then $-a$ is 0.
2. The opposite of $-a$ is a; that is, $-(-a) = a$.

It is not hard to imagine that native English-speaking students would have difficulty piecing together the logical statements in this section of text. And it is easy to see how non-native students might have even more difficulty when reading it in English.

Discussion of Semantic Features

Problems involving syntactic patterns are compounded by semantic phenomena ranging from the meanings of isolated vocabulary items (e.g., two-word verbs like *divided into*), to paraphrase relations between both natural language and formal language expressions (e.g., four divided into nine; nine divided by four; $9 \div 4$), to the referents of variables (e.g., $5g = b$ where g and b represent numbers of girls and boys), to inferences that may be drawn from linkages established by logical connectors (e.g., if . . . then sentences). Although students may have substantial practice with the syntactic patterns of English, there is no guarantee that they have acquired the denotative, connotative and conceptual network (the semantic patterns) that accompanies the individual words, phrases, and sentences common in mathematical discourse.

Lexical Items (Vocabulary).   Mathematics vocabulary includes a set of words that are specific to mathematics, which must be learned new by most students. These words—*divisor, denominator, quotient,* and *coefficient*—are relatively easy to learn. However, the mathematics register also includes everyday vocabulary items that have a different meaning in mathematics. Words such as *equal, rational, irrational, column,* and *table* have to be learned again, this time in a math context, since they have a specialized meaning in mathematics.

In addition to isolated vocabulary items, complex strings of words or phrases are used. If is often the case that the combination of two or more mathematical concepts are put together to form a new concept, thereby compounding the task of comprehending the words. The phrases, *least common multiple, negative exponent,* and even something apparently simple like *a quarter of the apples* are good examples of the complexity of mathematical phrases.

A subtler and much more difficult aspect of math vocabulary involves the many ways in which the same math operation can be signaled. As students progress through the hierarchy of math skills, manipulation of this vocabulary becomes crucial for understanding teacher explanations in class and for solving word problems. Crandall et al. (in press) have identified groups of lexical items in beginning algebra that signal that certain operations should be undertaken. For example, addition can be signaled by any of these words:

add	and
plus	sum
combine	increased by

Similarly, subtraction can be signaled by these words:

subtract from	minus
decreased by	differ
less	less than

It is important to note that the meanings of such terms are tied to specific operations. For example, students learn early in their mathematics education that the word *by* signals multiplication as in the expression *three multiplied by 10.* However, when later faced with algebraic expressions such as *a number increased by 10,* where *by* is part of an expression that signals addition, they are confused. Prepositions, in general, and the relationships they indicate are critical lexical items in the math register, items that cause a great deal of confusion. Additional examples include *divided by* versus *divided into* or *toward* and *to* in word problems involving distance (Crandall et al., in press; p. 11).

In addition to words and phrases particular to the math register, students must also learn the set of notational symbols used in expressing mathematical concepts and processes. The more advanced the mathematics, the greater the number of symbols and the more conceptually dense their meanings. Hence students seeing symbols such as $<$ and $>$ and parentheses () and brackets [] must learn how to relate symbol to mathematical concept or process (morst likely couched in math language) and then translate these into everyday language in order to express the mathematical ideas embedded in the symbols.

It is also important to remember that there are some symbols that have different meanings depending upon the part of the world in which they are used. In a number of Latin American countries, for example, a comma is used to separate whole numbers from decimals—a function of the decimal point in the United States—and the decimal point is used with whole numbers to separate hundreds from thousands, hundred thousands from millions, and so forth (e.g., 4.500,36 instead of 4,500.36).

Inferential Meaning and Reference.   Correctly manipulating the special vocabulary and phrases and the word order found in mathematics discourse is intricately tied to the ability to infer the correct mathematical meaning from the language. Making such inferences often depends on the language user's knowledge of how reference is indicated. In algebra, for example, the correct solution for word problems often hinges upon identifying key words and then knowing the other words in the problem to which those key words refer. For example, in a problem such as

Five times a number is six more than two times the number. Find the number.

students must realize that *a number* and *the number* refer to the same quantity.
In the problem,

> The product of two numbers is 90. If the first number is 10 times
> the other, find the number.

students must know that they are dealing with two numbers. Furthermore, they must know that the wording of the problem links each number with unique information; for example, the referent of *the first number* and the information given about it (*10 times the other*). Moreover, students must know that the translation from the words of the problem to the symbolic representation of the solution equation will be based on only one variable, and that each of the two numbers will be expressed in terms of that one variable.

A common mistake for students who realize the problem is about two numbers is to write solution equations with two *different* variables. They see no relationship between the two numbers described in the problem and consequently cannot write an equation using only one variable to represent the two numbers.

The above examples point to another referential feature of the language of mathematics: the identification of variables. Identifying the referents of variables is essential to correctly translating the words of a problem into the symbols of its solution equation. Variables stand for the *number* of persons or things, not for the persons or things themselves. The classic "students and professors" type of word problem illustrates this point:

> There are five times as many students as professors in the mathematics department.
> Write an equation that represents this statement. Use s to represent the number of
> students and p to represent the number of professors.

Many students write the following incorrect equation, $5s = p$, which follows the literal word order of the natural language sentence and uses s to represent "students" and p to represent "professors." The correct equation is: $5p = s$, which can be determined only if students know that the variable s (or any other variable they choose to use) must represent the NUMBER of students and that the variable p must represent the NUMBER of professors. This "reversal error" has been the subject of investigation by a number of researchers interested in the language of mathematics, including Clement (1982) and Mestre, Gerace, and Lochhead (1982). Firsching (1985) points out that recurring reversal errors could result from students' previous, repeated exposure in "beginning" level mathematics to word problems whose solution equations consistently require a translation based on a one-to-one correspondence between words and symbols. Intensively trained to solve word problems this way, they incorrectly assume that all problems can be solved in the same manner.

Discussion of Pragmatic Features

Syntactic patterns and semantic relationships occur in the context of larger stretches of discourse, i.e., the expository prose and exercises found in textbooks and the classroom lectures and handouts prepared by teachers. Thus, it becomes necessary to go beyond the strictly linguistic features of mathematics language to consider possible difficulties relating to extralinguistic contexts of utterance, i.e., the places, persons, and times involved as well as the beliefs, intentions, presuppositions, and background knowledge of participants in mathematical discourse. Below, we isolate a few of the epistemological problems noted in our research, saving discussion of the co-occurence of syntactic, semantic, and pragmatic features in math discourse for our analysis of protocols in the next section.

Students who lack certain kinds of experience, or whose experience has been different from or even contradictory to the experiences presupposed by certain word problems, are apt to encounter difficulties. For example, students who are unfamiliar with an economic system that encourages competition and private enterprise are likely to misunderstand such business concepts as discounts, mark-ups, wholesale, and retail. Or, if students happen to know only the tax rate in their locality, then it is not surprising that they suppose that rate applies everywhere, even if it is the unknown in the problem. Some students balk at the notion of solving for the tax rate because, in real life, the tax rate is always known and it is the amount of tax that needs to be calculated. However, because actual tax tables employ a rounding-off system that differs from the rule of thumb adopted by math texts, i.e., that we round off at .5 and above, there is the possibility of error even where the student has been encouraged to go into the marketplace to test his or her math skills.

The pedagogical advantages of incorporating real situations in an interactive framework, a common practice in most English as a second language (ESL) curricula, seem to be absent in the traditional mathematics curriculum. Communicative breakdowns are apt to occur when math texts and classroom lectures proceed in a rigid, lockstep manner, i.e., in a manner whereby the text or the teacher provides a statement plus explanation of a rule or property, demonstrates a few examples, and then gives the student problems to solve. It is often the case that these problems are unrelated to student experience, a problem noted above, which means that students have never had the opportunity to discuss the concepts that are involved.

A LINGUISTIC ANALYSIS OF SOME PROBLEM-SOLVING PROTOCOLS

In the previous section, we provided a linguistic model for categorizing the syntactic, semantic, and pragmatic features of the mathematics register. In this section, we focus on the interplay between these features that occurs when students are engaged in problem-solving tasks. Such a focus should make it clear that the three types of features are related in complex ways that make mastery of the mathematics register difficult and that necessitate the special kinds of teaching materials and techniques discussed in the next section.

Each of the two problems cited here was presented to groups of students in beginning algebra classes. The sessions, led by a researcher-facilitator, were taped, transcribed, and analyzed to identify linguistic difficulties. Most of the students were non-native speakers of English with varying degrees of English proficiency. A few of the students were, however, native English speakers. For both of the problems, portions of three different transcripts are used to illustrate a variety of syntactic (SYN), semantic (SEM), and pragmatic (PRAG) features that were problematic for the participating students.

Problem 1: The sales tax is $15 on the purchase of a diamond ring for $500. What is the sales tax?

(S1 = first student; S2 = second student; R = researcher)
Problem 1
Transcript 1

SEM   S1: That makes me confused sometimes in understanding *on the purchase of a diamond.*

R: Oh, so this phrase, *on the purchase of a diamond . . . ?*

S1: Right.

R: . . . confused you. Was it the term *on* that . . .

PRAG S1: Yeah, OK. I was, in my language what you sometimes have to do that, suppose if you purchase, like the purchase of 500, 500 dollars and sometimes we do like that way. It makes an understanding problem.

SEM   S1: . . . now I'm saying that the 15 *on* the 500 or 15 dollars included in the 500 like 485 plus 15 dollars is 500. My purchase is 485 so I would like to say that the prepositional phrase *on the,* like suppose a customer bought 500 dollars goods and he paid 15 dollars tax on that and what percent sales tax did he pay on 500 dollars. Like that way, you know?

Problem 1
Transcript 2

PRAG S1: Well, we know here in Miami it's 5%. So you have to divide by . . .

SEM   OK! 15 over 100, I mean 500. I don't know.

SYN   S2: Can I help? I suggest that you divide 500 by 15 and that will give you the rate.

S1: Right!
R: Tell me again. You divide 500 by...
S1: 15.
R: Let's do it and see what we get.
(S1 calculates the answer)
S1: OK. It's 3%.

Problem 1
Transcript 3

SEM   S2: The 15 dollars is the sales tax and the price of the ring is
            500, so it would be 515 dollars. But now how do I get the
            sales tax *rate?* What do I have to do? Divide?
R: Keep going.
PRAG   S2: What I'm thinking is . . . but then again, maybe it isn't
            plus 5%.

Discussion of Problem 1.   Excerpts from these three problem-solving sessions involving this same sales tax problem reveal student difficulties on the
syntactic, semantic, and pragmatic levels. In Transcript 1, the student admits
confusion regarding the phrase *on the purchase of a diamond ring for $500.* This
confusion might relate to the lexical meaning of the idiom *on the purchase of,* or
to the precise reference of *the purchase,* which can be $500 or $485. The student
chooses to disambiguate the problem by opting for an interpretation based on
experience with his native language (Bengali). That is, in his country, (Bangla-
desh) speaking his language, the total purchase would be $500 minus $15 rather
than $500 plus $15. Thus, this student exhibits difficulties related both to semantic (lexical) and pragmatic (epistemological) features.
     In Transcript 2, the attempt to solve the problem reveals that S1 does not
know how to translate the term *divide by* into math notation, i.e., doesn't know if
15 is the numerator or denominator. S2 makes a syntactic error in reversing the
order of the numerator and the denominator ("500 by 15" instead of "15 by
500"). Yet, S1 gets the correct answer, meaning that the correct order was
chosen despite the mistaken suggestion of S2. One wonders what S1 would have
done if S2's suggestion had been correct!
     In Transcript 2, S1 makes reference to the sales tax in Miami, as does S2 in
Transcript 3. In the first case, S1 seems to think that this bit of knowledge will be
helpful in setting up the solution, an idea that is quickly discarded. In the second
case, S2, after exhibiting some semantic confusion regarding a perceived difference between sales tax and sales tax rate, is unable to abandon the notion that
the answer is 5% despite the fact that dividing 15 by 500 does not support such a
solution.

Problem 2:   Two cars that are 375 miles apart and whose speeds differ by 5 miles per hour are moving toward each other. They will meet in 3 hours. What is the speed of each car?

Problem 2
Transcript 1
    PRAG S1:   The wording is what's difficult.
          R: What's bad about the wording?
    PRAG S1:   That one with the cars. Why can't they just say "One car . . . was driving x and the other car was driving $x + 5$?"
          R: You want them to set it up for you, huh? (Laughter)
        S1:   Right!

Problem 2
Transcript 2
          R: Rick, what are you doing over there. Help us out!
    PRAG S1:   OK. Well, for example, you have 2 cars. So I drew 2 cars (like blocks). And they're 375 miles apart, so I put 375 in the middle. "Whose speeds differ by 5 miles per hour." They're moving *toward each other*. They will meet in 3 hours. What is the speed of each car?" So what I'm beginning to do is I divided 375 by 2, so there's approximately 187 miles to the middle. Here's where I'm stuck. I am now trying to think up a formula. Maybe if I divide . . . 2 cars . . .

Problem 2
Transcript 3 (native English speakers)
        S2:   OK. Think out loud. Put the problem into context. 375 miles, um, whose speeds differ by 5 miles, so 50 or 55, 60 or 65. That's how they're going. One's going 5 miles an hour faster than the other, OK, let's assign a value to each car. Or for their speed. OK, so you got . . .
        S1: x and $x + 5$.

Discussion of Problem 2.   Problem 2 represents a type of problem widely reported to cause difficulties regardless of language proficiency. In Transcript 1, S1 complains about the wording, i.e., the textual difficulties involved in translating English into math notation, particularly in problems that do not relate to any obviously relevant real-life situation. In Transcript 2, S1 attempts to provide some reality by drawing a picture, but this only leads to an approximation, one that fails to come to grips with the relationship between distance, rate, and time necessary to solve the problem.

In Transcript 3, S1 and S2 are both native speakers of English who were instructed to follow a heuristic procedure similar to the four-step method promoted by Polya (1945). In this case, the attempt to put the problem into a context leads to an appeal to a piece of experience that enables them to approximate both the solution equation (x and $x + 5$) and the solution (60 or 65). Again, however, as in Transcript 2, they do not see any relationship between these numbers and variables and the distance, rate, time formula needed to set up the solution. In fact, a non-native English speaker who joined the discussion later did bring this important piece of mathematical knowledge with him, and this enabled the three students to work backwards from 60 and 65, x and $x + 5$, to the appropriate solution equation.

APPLICATIONS: A LANGUAGE APPROACH TO TEACHING MATHEMATICS

The previous examples illustrate the difficulties students face in attempting to acquire the mathematics register. They also suggest an approach to help students acquire this register: the development and implementation of interactive exercises and related teaching techniques that require students to discuss problems en route to solutions. The very act of talking through problems—of discussing various strategies for beginning the problem (or ''attacking'' it) and using the language to solve the problem—enable students to gradually become comfortable listening to and using mathematics language.

Although problem-solving sessions with students were initially used for data gathering, they also provided direction for an instructional approach. If students were given an opportunity to work together in a systematic way to develop competence in both listening and using math language, followed by practice in the written language, they could move beyond the stage of considering math language as a barrier, to seeing it as a tool for doing mathematics. That is, the language became an instrument that students could actively use, sometimes translating it into more familiar language terms. Moreover, they could benefit from mutual discussion and experience.

However, the process needed to be systematized to guide the practice and the acquisition. Thus, materials have been developed that provide students with an opportunity to understand math language in context and then to practice using it, both in explaining and actually solving the problems. Although the materials are intended to be used in paired tutoring sessions (and thus, are laid out with the tutor's page on the left and the comparable student's page on the right), they have also been used as classroom texts and as the basis for homework assignments. If they are used by students individually, the student is asked to look at the tutor's section for directions and verification. (An audio cassette of the tutor's role is being prepared for individual use with one of the units.)

The five units in these materials, *English Language Skills for Basic Algebra* (see Crandall et al., 1987), provide a review of basic mathematics and algebraic expressions, an introduction to equations and inequalities, an approach to and practice in solving a variety of word problems, an introduction to properties and theorems, and a glossary of key terms, symbols, types of numbers, and common phrases used in word problems. (The glossary is intended as a reference to be used throughout the program.) These units are cross-referenced to five college algebra textbooks used in the United States. The materials are carefully designed to help students move from an understanding of numerical and algebraic expressions, to the use of these in solving equations, inequalities, and word problems, and being able to understand definitions and theorems. For example, students might begin by being asked to understand that when their partners say "sixteen increased by eleven" that the appropriate numerical expression is $16 + 11$, just as it is if their partners say "sixteen plus eleven," "eleven added to sixteen," or "eleven more than sixteen." A subsequent exercise might introduce *square root* or an equation in which this same expression is required. As students progress through the materials, sometimes acting as tutors and other times as students, they become increasingly able to use math language to think about, discuss, and set up and solve problems. The unit on solving word problems also introduces a number of strategies for coping with the major difficulties in word problems: for example, finding ways of focusing on what is known (or unknown) and what is asked for, using strategies such as drawing diagrams to assist in the process; and watching out for key words such as *toward* (in mileage/distance/rate problems), *ago* or *will be* in age problems, and so forth.

Verbalization exercises can serve as a major component of class instruction or as a kind of supplementary tutorial or lab work. The materials have been used with the teacher and entire class in a traditional manner, in learning centers with two students working together, or in paired practice in class. They have been used in ESL classes as a means of increasing the academic content of English language classes and in algebra classes, as the basis for homework, small-group work, or even full class discussion.

A language approach to the teaching of mathematics provides multiple opportunities for students to develop listening, speaking, reading, and writing skills as they are acquiring mathematical skills. There are a number of other ways in which students can be assisted in developing language skills appropriate to their mathematics courses. For example, although the typical lesson in an algebra class focuses almost exclusively on the presentation of a problem type with sample demonstrations, followed by student practice, it is possible for teachers to provide more transition into problem solving by explaining the various linguistic items, giving students an opportunity to work together to solve problems (while the teacher listens for problematic language and math features), and asking discussion questions. These activities include:

- providing opportunities for students to "create" word problems, using numbers and other information provided by the teacher;
- preparing students for and then asking essay-type questions on exams, which force students to use math language rather than mechanically calculate solutions to problems;
- using dialogue-journals with students, providing students with an opportunity to write about their math problems, in a written dialogue between teacher and student about the nature of mathematics and the individual student's progress through the course. These journals also serve as a place to answer questions that the student may not have felt comfortable articulating in front of the entire class.

In summary, a language approach to mathematics education offers students the opportunity to acquire competence in both understanding and using mathematics language: in being able to understand instructor's presentations, text explanations, problems, and tests; in being able to ask questions and discuss problems in class; and in being able to work through algebraic, geometric, or other problems, using the mathematics language appropriate to the task. Math-anxious students who seem unable to respond when confronted by the mathematics register can be assisted in learning to cope with the language and even to effectively use that language in doing mathematics. However, this can only be accomplished when there is a conscious effort on the part of the instructor to provide activities that will facilitate the learning of this complex and difficult register. To expect students to "pick up" the language as they read the texts and listen to explanations in class is to court the kind of failure that too many students, both language minority and English-speaking, meet in their mathematics courses. The direction of our work, as applied linguists, and the work of some of our colleagues in the field of mathematics education, is to provide materials that will enable instructors to assist students in both the language and principles of mathematics.

ACKNOWLEDGMENTS

This research was supported by the Fund for the Improvement of Postsecondary Education, Grand No. G 00840473, 1984. The authors wish to thank the mathematics departments of Metropolitan State College, Denver, Colorado; Miami-Dade Community College, Miami, Florida; and Northern Virginia Community College, Alexandria, Virginia for their participation in the planning, research, and materials development stages.

REFERENCES

Aiken, L. R. (1971). Verbal factors and mathematics learning: A review of research. *Journal for Research in Mathematics Education, 2*(4), 304–312.
Burns, M., Gerace, W. J., Mestre, J. P., & Robinson, H. (1983). The current status of Hispanic

technical professionals: How can we improve recruitment and retention? *Integrated Education,* *20,* 49–55.

Carnap, R. (1955). Foundations of logic and mathematics. *International Encyclopedia of Unified* *Science, 1*(3), 139–212. Chicago, IL: University of Chicago Press.

Clement, J. (1982). Algebra word problem solutions: Thought processes underlying a common misconception. *Journal for Research in Mathematics Education, 13,* 16–30.

Cossio, M. G. (1978). The effects of language on mathematics placement scores in metropolitan colleges. *Dissertation Abstracts International, 38,* 4002A–4003A (University Microfilms No. 77–27,882)

Crandall, J., Dale, T. C., Rhodes, N. C., & Spanos, G. (1987). *English language skills for basic* *algebra.* Englewood Cliffs, NJ: Prentice-Hall.

Crandall, J., Dale, T. C., Rhodes, N. C., & Spanos, G. (in press). The language of mathematics: The English barrier. *Proceedings of the 1985 Delaware Symposium on Language Studies VII.* Newark, DE: University of Delaware Press.

Cuevas, G. (1984). Mathematical learning in English as a second language. *Journal for Research in* *Mathematics Education, 15,* 134–144.

Cummins, J. (1981). The role of primary language development in promoting educational success for language minority students. In California State Department of Education, Office of Bilingual Bicultural Education (Ed.), *Schooling and language minority students* (pp. 3–50). Los Angeles: Evaluation, Dissemination and Assessment Center, California State University.

Dawe, L. (1984, August). *A theoretical framework for the study of the effects of bilingualism on* *mathematics teaching and learning.* Paper presented at the Fifth International Congress on Mathematical Education, Adelaide, Australia.

DeAvila, E., & Havassy, B. (1974). *IQ tests and minority children.* Austin, TX: Dissemination Center for Bilingual Bicultural Education.

Dolciani, M. P., & Wooton, W. (1970). *Book one, modern algebra, revised edition.* Boston: Houghton Mifflin.

Duran, R. (1979). *Logical reasoning skills of Puerto Rican bilinguals* (Final Report Grant No. NIE-G-78—135). Washington, DC: National Institute of Education.

Firsching, J. T. (1985). *Conversational implicatures: Mathematics word problems and teacher* *explanations* (Vol. 1). Unpublished dissertation, Georgetown University, Washington, DC.

Gerace, W. J., & Mestre, J. P. (1983). *A study of the algebra acquisition of Hispanic and Anglo* *ninth graders* (Final Report on NIE Contract 400-81-0027). Amherst, MA: University of Massachusetts, Department of Physics and Astronomy. (ERIC ED 231613)

Halliday, M. A. K. (1975). Some aspects of sociolinguistics. In E. Jacobson (Ed.), *Interactions* *between linguistics and mathematics education* (pp. 64–73). *Final Report of the Symposium* *Sponsored by UNESCO, CEDO and ICMI,* Nairobi, Kenya, September 1–11, 1974. (UNESCO Report No. ED-74/CONF. 808. Paris: UNESCO, 64–73)

Kessler, C., Quinn, M. E., & Hayes, C. W. (1986). Processing mathematics in a second language: Problems for LEP children. To appear in *Proceedings of the 1985 Delaware Symposium on* *Language Studies VII.* Newark, DE: University of Delaware Press.

Knight, L., & Hargis, C. (1977). Math language ability: Its relationship to reading in math. *Language Arts, 54,* 423–428.

Mestre, J. P. (1981). Predicting academic achievement among bilingual Hispanic college technical students. *Educational and Psychological Measurement, 41,* 1255–1264.

Mestre, J. P. (1986). Teaching problem-solving strategies to bilingual students: What do research results tell us? *International Journal of Mathematical Education in Science and Technology, 17,* 393–401.

Mestre, J. P., Gerace, W. J., & Lochhead, J. (1982). The interdependence of language and translational math skills among bilingual Hispanic engineering students. *Journal of Research in Science* *Teaching, 19,* 399–410.

Moreno, S. (1970). Problems related to present testing instruments. *El Grito, 3,* 25–29.

Morris, C. W. (1955). Foundations of the theory of signs. *International Encyclopedia of Unified Science. 1*(2) 78–137. Chicago, IL: University of Chicago Press.

Munro, J. (1979). Language abilities and math performance. *Reading Teacher, 32,* 900–915.

Polya, G. (1945). *How to solve it.* Princeton, NJ: Princeton University Press.

Ramirez, M., & Gonzalez, A. (1972). Mexican Americans and intelligence training. In M. M. Mangold (Ed.), *La causa Chicana: The movement for justice* (pp. 137–147). New York: Family Services Association of America.

Thorndike, E. L. (1912). The measurement of educational products. *School Review,* 20, 289–299.

Chapter 14

Bilinguals' Logical Reasoning Aptitude:
A Construct Validity Study

Richard P. Durán

Graduate School of Education
University of California
Santa Barbara, CA

BILINGUALISM AND COGNITION

How does the proficiency in a language affect the ability to solve problems? This is an important question for educators, testing and assessment specialists, and cognitive psychologists. Educators are interested in better understanding ways in which non-English background students can function effectively in classrooms offering instruction in one language or another. They need to understand whether the choice of a language of learning and problem solving will affect display of the underlying cognitive skills of bilingual students. Educators are especially interested in understanding how students' earlier schooling experiences and backgrounds are related to their language proficiency skills, as well as to the cognitive skills in which students have developed competence.

Testing specialists in the schools have related concerns. They are often called upon by teachers and school staff to answer questions about how to best serve bilingual students; it is often the burden of testing specialists to assess the cognitive and language proficiency skills of bilingual students. Further, when trained appropriately, testing specialists can also help contribute to explanations of how bilingual students' backgrounds might be related to their learning ability in either language.

Cognitive psychologists are more removed from the everyday contexts of schools. However, they too are interested in some of these questions. They are interested in developing scientific theories of how the mind functions when operating in one language or another. Like educators, cognitive psychologists benefit from the research of testing specialists. The findings uncovered through testing research provides empirical evidence of bilinguals' ability to function in two languages. By considering this evidence, cognitive psychologists can develop theories and conduct research providing scientifically sophisticated theories of bilinguals' cognitive functioning than is possible by use of tests alone.

This chapter discusses research on Puerto Rican adult bilinguals' logical reasoning skill in Spanish and English. It is written from the perspective of testing and assessment. The research described here bridges the concern of educators and cognitive psychologists. It examines similarities and differences in bilinguals' performance on logical reasoning tests in each language. Further, it examines how performance on reasoning tests is related to measures of students' reading proficiency in each language and to students' schooling in Puerto Rico and parents' socioeconomic background.

Previous Research

The principal question addressed by the research described here is whether bilinguals performed similarly on tests of logical reasoning that differ only in the language used to state reasoning problems, given information about their reading ability in each language. Surprisingly there has not been much research on bilinguals' high-level reasoning skills despite the importance of such skills to academic learning. Instances of such research do exist, however. For example, John Macnamara (Macnamara & Kellaghan, 1968, cited in Macnamara, 1967) found that Irish bilinguals were more likely to make errors in solving verbal mathematics problems in their non-native language, Irish, than in their native language, English. He found evidence that some of the bilinguals could understand the individual sentences making up verbal mathematics problems in Irish, and that they also knew the mathematics operations required for solving problems, but that they proved unable to solve problems, nonetheless. A more recent example of research on higher level thinking abilities of bilinguals is provided by d'Anglejan, Gagnon, Hafez, Tucker, and Winsberg (1979). These investigators found that Canadian bilinguals were less accurate in solving syllogisms in their second language, English, than in their first language, French. The subjects were equally likely to understand syllogism premises in either language, but they were poorer at solving problems in their second language. The data also indicated that bilinguals showed more variability in the speed with which they worked problems in their second language as compared to their first language.

Mestre (1984) compared the performance of Anglo and Hispanic college students interpreting complex premises stated in English. Mestre found that Hispanic students were significantly slower in the task than were Anglo students. Significant differences, however, were not found in the accuracy of solutions across Anglo and Hispanic groups. Mestre suggested that slower rate of problem solution among Hispanics could stem from their limited familiarity with English as compared to Anglo students. This study did not uncover evidence of linguistic interference effects among Hispanics, however. For instance, lack of parallelism in how negative constructions are expressed across the two languages did not lead to depressed performance of Hispanics on premise comprehension. It may be that accuracy, in general, increases when solution time is extended, and perhaps the language issue for Hispanics became one of *facilitation* of accuracy with increased time spent on the problem.

In another recent study, Durán and Enright (1983) found little difference in Hispanic bilingual college students' ability to solve matching syllogism problems in Spanish and English. The subjects in the study had greater verbal proficiency in English than in Spanish. Durán and Enright found a correlation of .96 between students' accuracy of solution of the same syllogism problems across the two languages. The results also indicated that students solved problems more slowly in Spanish—their least proficient language—than in English, though there was a substantial correlation in the speed with which subjects solved problems in the two languages.

The study described in this chapter contributes to the foregoing studies by focusing attention on bilinguals' ability to solve problems on several matched pairs of Spanish and English reasoning tests. Research is not clear on whether bilinguals are equally capable of solving problems in either language and whether performance on tests is separable from language proficiency in either language. The present study used the technique of confirmatory factor analysis to investigate this issue. This technique allows a researcher to test statistically hypotheses about ways in which tests of different cognitive and linguistic abilities are interrelated to each other given the intercorrelations that they exhibit. Thus, it was possible in the present study to investigate whether the intercorrelations among reasoning and reading test scores constituted separable factors that could be identified. Additional analyses were undertaken in the study to learn how background characteristics of subjects associated with cognitive and linguistic abilities.

SUBJECTS AND MEASURES

Subjects

The subjects investigated were 104 Puerto Rican adults enrolled in 21 4-year and 2-year colleges on the East Coast in 1978-1979. Over 80% of the subjects were enrolled in 4-year colleges. Schools varied in their academic orientation, with several Ivy League as well as small state and private colleges represented. About one half of all subjects had been born in Puerto Rico, the remainder in the United States. All subjects were screened at the time of their participation for the presence of a bilingual background and adequate knowledge of Spanish to understand the requirements of Spanish version tests. Persons with no proficiency in Spanish were excluded from the study. Subjects also varied in their English proficiency, but all subjects were sufficiently proficient in English to function in English-only college classrooms. The original number of subjects investigated was 209. However, subjects with missing data were excluded from analysis in the present chapter.

Logical Reasoning Measures

Scores on three logical reasoning tests in Spanish and English versions were analyzed. These tests and their appropriate abbreviations are Diagramming Relationships (DIAGRELS); Inference Test (INFER); Logical Reasoning (LOG-REAS). A fourth test was administered, Nonsense Syllogisms, which is excluded from consideration because of its low internal consistency reliability. A brief description of each of the tests is given in Appendix A, drawn from Duran (1979).

English language versions of the tests used in this study had been found to measure a common set of skills that were labeled *logical reasoning* (Ekstrom, French, & Harman, 1976; French, Ekstrom, & Price, 1963). Procedures used in creating Spanish versions of the test along with reliability characteristics of the tests are described in Duran (1979). Two native-speaking Spanish translators were used to create the Spanish version tests. One of the translators was a Puerto Rican sociolinguist who evaluated all translations for appropriateness given the background and schooling characteristics of the Puerto Rican population to be studied. One-half length tests were used, based on the original full-length English version of the tests. One of the two halves of a given test was administered in its English version, and the other half was administered in its Spanish version. In the study, use of each language version of a one-half length test was counterbalanced so that approximately the same number of subjects were administered each half-length version of a test in each language. Order of language for first presentation of a test was counterbalanced with approximately one half of all subjects receiving a test in one language before a test in the other language. Durán (1979) describes other language proficiency data collected during the course of administering reasoning tests.

Reading Comprehension Measures

Two pairs of advanced reading comprehension tests in Spanish and English were utilized. These tests were the *Prueba de Lectura, Nivel 5 - Advanzado - Forma DEs* and the *Test of Reading, Level 5 - Advanced Form - CE* (Guidance Testing Associates, 1962). Both instruments were developed together and were intended to be highly parallel in form, content, and difficulty (Manuel, 1963). Each test has three sections and subscores: Vocabulary, Speed, and Level. For each language the corresponding abbreviations are VOCAB-Sp or VOCAB-Eng; SPEED-Sp or SPEED-Eng, and LEVEL-Eng or LEVEL-Sp. The Vocabulary subtest measured recognition vocabulary, the Speed subtest measured ability to answer correctly numerous simple questions about a passage in a brief period of time, and the Level subtest measured ability to recognize paraphrases and conclusions drawn from item passages that were paragraph length.

Background and Personal Measures

The measures considered for analysis included Years Lived in Puerto Rico (PRYRS); Years Lived in the U.S. Mainland (USYRS); Years Schooled in Puerto Rico (PRSCH); Years Schooled on the U.S. Mainland (USSCH); Prestige of Planned Occupation (ASPIRNORC); Prestige of Father's Occupation (FATHNORC) and Prestige of Mother's Occupation (MOTHNORC). Information on these measures was derived from a background questionnaire administered to subjects. Prestige of occupation measures were based on the NORC scale prestige rating for the professions indicated.

RESULTS

Intercorrelations Among Measures and Factor Models

Table 14.1 presents intercorrelations among all measures. Inspection of this table is helpful because it shows how pairs of variables are interrelated. Inspection of Table 14.1 reveals considerable intercorrelation among reasoning measures and language proficiency measures, regardless of language. This was expected given findings of previous research on the association between measures of general aptitude and language proficiency test scores. Among background measures, there appeared to be a nearly perfect correlation among variables reflecting years schooled and lived in Puerto Rico versus the United States. This, too, was expected, given the dependence among these measures. In subsequent factor analyses, only years schooled in Puerto Rico was retained in the analysis. Prestige of students' planned occupation showed higher intercorrelations with measures of logical reasoning and reading comprehension proficiency than did prestige of mother's or father's occupation.

Confirmatory factor analysis procedures were used to investigate the two questions under research. Confirmatory procedures allow an investigator to specify mathematical models indicating how variables may contribute to measurement of underlying factors (i.e., constructs) of interest. Each model specified is a "theory," or set of hypotheses, which the investigator believes can account for the patterns of intercorrelations observed among variables. The investigator then

TABLE 14.1
Correlation Matrix

	DIAGRELS-Sp	INFER-Sp	LOGREAS-Sp	DIAGRELS-Eng	INFER-Eng	LOGREAS-Eng	VOCAB-Sp	SPEED-Sp	LEVEL-Sp
DIAGRELS-Sp		.564	.543	.580	.489	.586	.541	.503	.548
INFER-Sp			.490	.420	.491	.510	.521	.526	.489
LOGREAS-Sp				.474	.420	.728	.513	.509	.454
DIAGRELS-Eng					.515	.563	.349	.428	.480
INFER-Eng						.488	.484	.454	.481
LOGREAS-Eng							.382	.425	.437
VOCAB-Sp								.739	.744
SPEED-Sp									.771
LEVEL-Sp									
VOCAB-Eng									
SPEED-Eng									
LEVEL-Eng									
PRYEARS									
USYEARS									
PRSCH									
USSCH									
ASPIRNORC									
FATHNORC									
MOTHNORC									

	VOCAB-Eng	SPEED-Eng	LEVEL-Eng	PRYEARS	USYEARS	PRSCH	USSCH	ASPIRNORC	FATHNORC	MOTHNORC
DIAGRELS-Sp	.496	.534	.584	.081	-.160	.153	-.146	.281	.271	.148
INFER-Sp	.510	.565	.652	.047	-.114	.117	-.114	.167	.135	.136
LOGREAS-Sp	.405	.418	.453	.169	-.255	.255	-.253	.271	.138	-.014
DIAGRELS-Eng	.465	.540	.598	-.090	-.052	-.088	.090	.217	.150	.154
INFER-Eng	.678	.635	.649	-.026	.007	-.004	-.000	.336	.225	.155
LOGREAS-Eng	.443	.491	.569	-.063	-.072	-.008	.014	.350	.059	.030
VOCAB-Sp	.693	.592	.550	.446	-.341	.460	-.461	.222	.194	.146
SPEED-Sp	.509	.621	.522	.480	-.457	.516	-.501	.137	.192	.072
LEVEL-Sp	.577	.609	.587	.315	-.400	.361	-.338	.240	.372	.163
VOCAB-Eng		.789	.734	.025	.039	.013	-.021	.374	.209	.218
SPEED-Eng			.771	.027	-.066	.034	-.026	.285	.187	.068
LEVEL-Eng				-.040	-.063	-.025	.028	.275	.179	.046
PRYEARS					-.842	.943	-.945	.014	.210	-.086
USYEARS						-.846	.834	-.060	-.339	.047
PRSCH							-.996	.063	.247	-.067
USSCH								-.055	-.234	.071
ASPIRNORC									.241	.200
FATHNORC										.296
MOTHNORC										

uses a statistical computer program, such as LISREL, to estimate each model. The output for each model indicates the loadings (i.e., weight or numerical importance) of each variable in predicting each of the factors specified by the model. The output also yields a numerical goodness of fit index for each model; the goodness of fit index tells the investigator how well a model can explain the observed intercorrelations among variables. The higher the index value, the better the model fits the data. An additional step in the analyses is to compare the fit of the various models to each other. By conducting appropriate statistical analyses, the investigator is able to compare the improvement in fit among models, e.g., going from the model with the simplest hypotheses to the models with the most complicated hypotheses. Subsequently, the investigator is able to pick the most parsimonious model that best accounts for the intercorrelations among variables.

The strategy for conducting factor analyses proceeded in two stages. In the first stage, the relative fit of three confirmatory factor models (A, B, and C) involving only logical reasoning and reading comprehension measures was evaluated using the LISREL program (Joreskog & Sorbom, 1981). The first stage analyses addressed the question of whether or not the reasoning and reading comprehension measures in each language were measuring the same constructs. Each model (A, B, and C) specified different hypotheses about the interrelationships among reasoning and comprehension measures in each language. Subsequently, in the second stage, extension loadings were computed using LISREL between background measures and factors for Model C. This analysis was based on the best fitting model obtained from stage 1. The purpose of the analysis was to learn how well reasoning and reading factors could be predicted from information about the number of years students had been schooled in Puerto Rico, the prestige of the students' job aspirations, and the prestige of mothers' and fathers' occupations.

Stage One: Explaining Relationships Among Reasoning and
Reading Measures

Factor Model A postulated that only a single general factor was needed to account for interrelationships among reasoning and comprehension measures. Model B postulated two correlated factors; one factor was a logical reasoning factor loading on Spanish and English reasoning test scores and a second factor loaded on reading comprehension scores on either Spanish or English. A final model, Model C, postulated three correlated factors. It was identical to Model B except that it postulated separate but correlated reading comprehension factors in each language.

Table 14.2 displays the value of the goodness of fit index (AGFI) obtained from LISREL confirmatory factor analyses for Models A, B, and C. Model C fits the data best, followed by Models B and A, respectively. Discussion of the sensibility of the models is addressed as each one is discussed in turn.

TABLE 14.2
Goodness of Fit of Factor Models*

Model	Fit Index AGFI	RMS
A	.588	0.79
B	.600	0.73
C	.702	0.59

*The meaning of the AGFI index is discussed in Joreskog and Sorbom (1981, section I.I2). RMS stands for "root mean square residual"; it is the square root of the variance associated with a model.

Model A:	One-factor model
Model B:	Two-factor model; correlated reasoning and reading comprehension facors.
Model C:	Three-factor model; one general reasoning factor, one Spanish reading comprehension factor, one English comprehension factor. All factors correlated.

Table 14.3 displays the standardized factor loadings obtained for the one factor model, Model A. The standardized factor loadings indicate how important each variable is to prediction of a factor; in this case of course, we are concerned with only one factor. These loadings are estimates of the true loadings that would be obtained if we had data on an entire population rather than just a sample. The highest possible value for a loading is 1.0, and the smallest possible value is -1.0. Variables that show values near 1.0 or -1.0 are important contributors to prediction of the factor in question. In contrast, a loading of 0.0 would indicate that a variable does not contribute to prediction of a factor. A statistic, T, is computed for each of the loadings. If the value of this statistic is larger than about 2.0, it means that the true value of the loading in the population is highly unlikely to be 0.0.

All of the standardized factor loadings, except the first one that was fixed in the initial parameter specifications, are significantly different from zero, based on their corresponding T-values. The pattern of loadings for Model A seems to place slightly more importance on reading comprehension measures in English and Spanish than on logical reasoning measures in either language.

Table 14.4 displays the factor loadings and intercorrelation among factors for Model B, which postulated correlated reasoning and reading comprehension factors. Each factor was permitted to load only on a measure of the same type in either language. That is to say, measures of logical reasoning in each language were permitted to load only on an underlying logical reasoning factor common to

TABLE 14.3
Standardized Factor Loadings:
Model A

Measure	Factor	Loadings[a]
DIAGRELS-Sp	.709	
INFER-Sp	.702	(6.901)
LOGREAS-Sp	.620	(6.101)
DIAGRELS-Eng	.651	(6.402)
INFER-Eng	.724	(7.115)
LOGREAS-Eng	.654	(6.429)
VOCAB-Sp	.760	(7.464)
SPEED-Sp	.736	(7.233)
LEVEL-Sp	.762	(7.492)
VOCAB-Eng	.814	(7.994)
SPEED-Eng	.843	(8.272)
LEVEL-Eng	.843	(8.271)

[a]T-values are indicated in parentheses. No T-value is given for the first loading since its unstandardized form was fixed at the value 1.0.

TABLE 14.4
Factor Loadings and Intercorrelations Among Factors: Model B

	Factor Loadings[a]			
	Factor 1		Factor 2	
Measure	Logical Reasoning		Reading Comprehension	
DIAGRELS-Sp	.769	(8.915)	0.0	
INFER-Sp	.709	(7.971)	0.0	
LOGREAS-Sp	.715	(8.057)	0.0	
DIAGRELS-Eng	.701	(7.849)	0.0	
INFER-Eng	.701	(7.841)	0.0	
LOGREAS-Eng	.761	(8.793)	0.0	
VOCAB-Sp	0.0		.788	(9.398)
SPEED-Sp	0.0		.754	(8.823)
LEVEL-Sp	0.0		.783	(9.312)
VOCAB-Eng	0.0		.832	(10.212)
SPEED-Eng	0.0		.856	(10.681)
LEVEL-Eng	0.0		.827	(10.102)

Intercorrelations Among Factors

	Factor 1	Factor 2
Factor 1	1.0	.858
Factor 2		1.0

[a]T-values for loadings are indicated in parentheses.

both languages, and measures of reading comprehension were permitted to load only on an underlying reading comprehension factor common to both languages.

The results for Model B indicate that measures manifested loadings that made theoretical sense. The pattern of loadings of reasoning measures with the reasoning factor is highly similar, ranging from .701 to .769. The pattern of loadings of reading comprehension measures with the reading comprehension factor is also highly similar, but with slightly higher loadings occurring for English measures. The correlation between the reasoning factor and the reading comprehension factor was .858, suggesting a great deal of overlap across the two factors. The T-values associated with factor loadings indicate that all loadings are significantly different from zero.

Table 14.5 displays the factor loadings and intercorrelations among factors for Model C. This model postulated a single reasoning factor and separate reading comprehension factors in each language. The pattern of loadings of reasoning measures in each language with the reasoning factor was virtually the same as obtained with Model B. Spanish and English reading measures had high loadings with their corresponding language-based reading comprehension factor. These loadings tended to be slightly higher than had been obtained with Model B. Within a language, the magnitude of loadings was highly similar. In the case of all three factors, the T-values associated with loadings on measures indicated that all loadings were significantly different from zero.

TABLE 14.5
Factor Loadings and Intercorrelations Among Factors: Model C

Measure	Factor 1 Logical Reasoning		Factor 2 Sp. Reading Comprehension		Factor 3 Eng. Reading Comprehension	
DIAGRELS-Sp	.766	(8.878)	0.0		0.0	
INFER-Sp	.710	(7.981)	0.0		0.0	
LOGREAS-Sp	.710	(7.975)	0.0		0.0	
DIAGRELS-Eng	.704	(7.895)	0.0		0.0	
INFER-Eng	.705	(7.907)	0.0		0.0	
LOGREAS-Eng	.761	(8.786)	0.0		0.0	
VOCAB-Sp	0.0		.855	(10.534)	0.0	
SPEED-Sp	0.0		.865	(10.732)	0.0	
LEVEL-Sp	0.0		.881	(11.042)	0.0	
VOCAB-Eng	0.0		0.0		.858	(10.659)
SPEED-Eng	0.0		0.0		.896	(11.428)
LEVEL-Eng	0.0		0.0		.870	(10.898)

	Intercorrelations Among Factors		
	Factor 1	Factor 2	Factor 3
Factor 1	1.0	.750	.839
Factor 2		1.0	.770
Factor 3			1.0

^{a}T-values for loadings are indicated in parentheses.

The logical reasoning factor had a correlation of .750 with the Spanish reading comprehension factor and a correlation of .839 with the English reading comprehension factor. These results indicate a high degree of overlap between reasoning and reading comprehension factors, particularly in the case involving the English reading comprehension and the reasoning factor. The Spanish reading comprehension factor correlated .770 with the English reading comprehension factor. This correlation is quite high, but nonetheless suggests that the two factors measured can be discriminated from each other.

The goodness of fit statistics for Model C shown in Table 14.1 indicate that Model C fit the data much better than Models B and A. The improvement in fit between Models C and B was noticeably greater than the improvement in fit between Models B and A. The conclusion that can be drawn is that Model C fits the data best. That is to say that the measures analyzed exhibit a common logical reasoning factor across languages of assessment and separate reading comprehension factors in each language. Acceptance of this model, however, rests on accepting tbe conclusion that the three factors that were extracted are highly correlated with one another. Further evidence on the plausibility of Model C is acquired by interpreting how various background measures associate discriminably with each other.

Stage Two: Background Measures as Predictors of Reasoning and Reading Factors

The LISREL program was used to estimate loadings for Model C on the following measures: number of years schooled in Puerto Rico, prestige of student's planned occupation, father's occupational prestige, and mother's occupational prestige.

The statistical technique that was followed, known as factor extension, involved fixing the loadings of reasoning and reading comprehension measures on factors according to their maximum likelihood estimates obtained in the analysis for Model C. The four background measures were then allowed to load freely on each factor determined by a model. The LISREL program was used to compute maximum likelihood estimates of these latter loadings. Subsequently, the correlations between each background measure and a factor were computed.

Table 14.6 shows these correlations. Correlations significant at the $p < .05$ and $p < .005$ levels are indicated. Prestige of student's planned occupation (ASPIRNORC) showed a significant relationship to the logical reasoning factor and to the English reading comprehension factor. Prestige of student's planned occupation also manifested a significant correlation with the Spanish reading comprehension factor, but the relationship was lower than were the cases with each of the other two factors.

Prestige of father's occupation (FATHNORC) showed a statistically significant correlation with all three factors of Model C. The relationship, however,

TABLE 14.6
Correlations of Background Measures with Factors of Model C

	FACTOR		
Background Measure	Logical Reasoning	Sp. Reading Comprehension	Eng. Reading Comprehension
PRSCH	.104	.462**	.026
ASPIRNORC	.384**	.258*	.379**
FATHNORC	.264*	.370**	.295*
MOTNORC	.175	.058	.138

*p < .05
**p < .005

seemed to be noticeably higher with the Spanish reading comprehension factor than with the logical reasoning or the English reading comprehension factor. Mother's occupational prestige (MOTNORC) did not show a significant correlation with any of the three factors in Model C. Number of years schooled in Puerto Rico (PRSCH) showed a significant correlation with the Spanish reading comprehension factor, but virtually no relationship with the logical reasoning and English reading comprehension factors.

The results of the foregoing analysis indicate that the Spanish reading comprehension factor was discriminable from a general logical reasoning factor and an English reading comprehension factor. The support arises because the magnitude of correlations between background measures and the Spanish comprehension factor contrasted noticeably when compared to the correlations between background measures and the two other factors. In contrast to this finding, however, there did not appear to be any major differences in the patterns of intercorrelation between background measures and the logical reasoning factor and the English reading comprehension factor. This outcome supports the possibility that the logical reasoning factor and English reading comprehension factor are measuring skills that are associated in common with the background factors.

SUMMARY AND DISCUSSION

Reasoning and Reading

The factor analyses that were conducted support the conclusion that students exercised the same underlying logical reasoning skills in working reasoning tests presented in Spanish and English. This finding lends support to the conclusion that bilingual students may be capable of transfering high-level thinking skills across problems stated in either of their two languages. The bilingual students who performed well on reasoning tests in one language tended to perform well on similar tests in their other language. Duran (1979) found that these same students performed poorly on reasoning tests in the language in which they reported least proficiency. These results imply that students' ability to demonstrate reasoning skills is assessed best in the language they are most proficient in, but that high-level reasoning skills involve manipulations of information that are independent of language skills to some degree. This conclusion is consistent with the findings of earlier work on bilinguals' cognition cited in the introduction to this chapter.

An additional factor analysis finding was that students' reading comprehension scores in Spanish and English constituted another set of two factors, separate but related to a reasoning factor. It should be noted that all three factors in the best fitting factor model were correlated highly to each other. Students as a whole were more proficient in English than in Spanish (Durán, 1981), and indeed, the results showed that the English reading comprehension factor correlated more highly with the reasoning factor than did the Spanish reading comprehension factor. Thus, although all three factors were statistically distinguishable from each other, there was a fair amount of overlap in the skills represented across factors. This overlap may have resulted in part, for example, because all of the tests administered were multiple-choice tests. Skills in working multiple-choice items might have been shared across tests regardless of whether tests were intended to measure logical reasoning as opposed to reading comprehension skills. Some investigators might lend a different explanation for the correlation among the three factors. Oller (1979), for example, contends that language proficiency test scores reflect the same skills measured by tests of general cognitive abilities. In Oller's account this association is not an artifact of the method of testing, but rather the result of the fact that language and reasoning skills are innately and closely connected.

The difficulty in separating reasoning and reading comprehension factors reported here has educational implications and implications for further cognitive research. In everyday educational contexts it may prove very difficult to assess some high-level cognitive skills independently from language proficiency skills. The results reported here and the findings of previous research on bilinguals' cognitive abilities, however, do support the conclusion that it is best to assess cognitive abilities in the most proficient language (Durán, 1985). More refined research is needed on the cognitive processes employed by bilinguals in solving problems in each of their two languages. This research needs to explore carefully how language proficiency constrains the utilization of specific skills that underlie performance on cognitive ability tests. In addition, and in another direction, cognitive research is needed on how the demands of everyday academic performance affect bilinguals' display of logical reasoning and reading comprehension skills.

Background Factors, Reasoning, and Reading

Number of years schooled in Puerto Rico and prestige of father's occupation showed a higher relationship to Spanish reading comprehension skills than they did to logical reasoning skills or English reading comprehension skills. Number of years schooled in Puerto Rico showed a near-zero relationship to logical reasoning skills and to English reading comprehension skills. The latter finding indicated that the students sampled would not be expected to show a differentiation in their ability to participate in English language academic settings as a function of the extent of their Puerto Rican schooling. This is an important finding. It suggests that having been schooled in a non-English language (with some exposure to English instruction, as is the case in Puerto Rico) does not affect the acquisition of critical thinking skills appropriate for college.

Analyses, however, did show that students with higher occupational aspirations and with higher prestige of father's occupation were significantly advantaged in their logical reasoning skills and English reading skills over other students. A similar relationship was found for Spanish reading comprehension skills, though number of years of Puerto Rican schooling was also implicated as an important variable in this instance. Although more careful partial correlation analyses are needed, the results suggest the hypothesis that aspirations and family socioeconomic factors are more likely to affect logical reasoning ability and English reading comprehension ability than mere exposure to a predominantly English or Spanish environment prior to college.

The importance of number of years schooled in Puerto Rico to Spanish reading comprehension skills is an obvious one given that Spanish is the principal medium of instruction in Puerto Rico; instruction in Spanish in the United States was virtually nonexistent among the students investigated. In appreciating the fact that years of Puerto Rican schooling showed no association to English reading comprehension skills, one must remember that all Puerto Rican students are exposed to English by law within the public Puerto Rican school system. Within the private schools of Puerto Rico, English is used even more than in public schools.

The finding that prestige of mother's occupation showed virtually no association to reasoning and reading comprehension factors probably reflects social and cultural values that predominate. It is a reflection of the social role played by mothers as homemakers and by the restricted range of lower level occupations that they are able to attain. One can speculate that as women's participation in professional and white collar roles increases, their occupation prestige level will associate more strongly with their children's cognitive, linguistic, and aspiration characteristics.

Findings of the sort described above stop far short of generating satisfactory explanations of how the background and schooling characteristics of bilingual students affect their development of cognitive and linguistic skills. The results reported, however, are consistent with more detailed accounts of social and cultural mechanisms that may affect the cognitive and educational aspiration development of Hispanic children and older students. Laosa (1982), for example, describes a detailed model, supported by data, describing how parental education and socioeconomic level, parental interaction styles with children, and home-school compatibility characteristics affect the cognitive development and educational attainment of Hispanic children. Laosa (1984) also has presented evidence that the cognitive ability test scores of Hispanic and Anglo children as young as 2 years, 6 months of age already reveal associations with parents' socioeconomic level. This research supports the importance of considering how the cognitive skills of older bilingual students are related to their social background and family upbringing.

APPENDIX
Logical Reasoning Tests

The logical reasoning test with minimal linguistic processing requirements was the Diagramming Relations test. This test presented a subject with sets of three nouns such as:

dogs, mice, animals

which then had to be matched against one of three diagrams of the form:

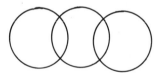 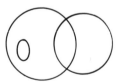

That captured the relationship among the conceptual categories referred to by words. In the case of the problem given, the middle diagram is the correct solution since dogs and mice are both animals but are distinct from each other. In terms of linguistic processing this test thus required only recognition of isolated words.

The Logical Reasoning, Form A test presented subjects with syllogisms that had to be classified as valid or invalid. Linguistically, this test required subjects to understand complete sentences and connections among simple descriptions across sentences.

The third logical reasoning instrument, know as the Inference Test, presented subjects with statements of the form:

All human beings fall into four main groups according to the composition of their blood: O, A, B, and AB. Knowledge of these blood types is important for transfusions.

On the basis of these statements subjects were required to select the unique correct conclusion that logically followed from a set of alternatives such as:

1. The blood type is determined by genes.
2. Persons of group AB can receive blood from any other type.
3. Blood transfusions between members of the same group are always safe.
4. Certain percentages of all people belong to each type. (correct answer)
5. Blood from persons of group O can safely be given to persons of any group.

Subjects were instructed to choose only that conclusion which followed from the original information given without bringing in other knowledge or beliefs not made explicit originally. Linguistically, this class of logical reasoning problems required the most extensive discourse comprehension skills of the four tests. Items on this test were the most complicated syntactically; they also required that subjects recognize pronoun and several other forms of reference within and across sentences.

ACKNOWLEDGMENTS

I would like to express my gratitude to Donald Rock and Leta Davis for their assistance in conducting the data analyses. This chapter was developed under National Institute of Education Contract 400-82-55008. Data analyzed in this chapter were originally collected under National Institute of Education grant NIE-6-78-0135. The views expressed in this chapter are those of the author and do not necessarily reflect the policy, opinion, or endorsement of the National Institute of Education.

REFERENCES

d'Anglejan, A., Gagnon, N., Hafez, M., Tucker, G. R., Winsberg, S. (1979). *Solving problems in deductive reasoning: Three experimental studies of adult second language learners.* Working paper on Bilingualism, No. 17, Montreal: University of Montreal.

Durán, R. P. (1979). *Logical reasoning skills of Puerto Rican bilinguals.* Final report to the National Institute of Education. Princeton, NJ: Educational Testing Service.

Durán, R. P. (1981). Reading comprehension and the verbal deductive reasoning of bilinguals. In R. P. Duran (Ed.), *Latino language and communicative behavior.* Norwood, NJ: Ablex.

Durán, R. P. (1985). Influences of language skills on bilinguals' problem solving. In S. Chipman, J. Segal, & R. Glaser (Eds.), *Thinking and learning skills; Vol. 2. Current research and open questions.* Hillsdale, NJ: Lawrence Erlbaum Associates.

Durán, R. P. & Enright, M. (1983). *Reading comprehension proficiency, cognitive processing mechanisms, and deductive reasoning in bilinguals* (Final report to the National Institute of Education). Princeton, NJ: Educational Testing Service.

Ekstrom, R. B., French, J. W., & Harman, H. H. (1976). *Manual for kit of reference tests for cognitive factors*. Princeton, NJ: Educational Testing Service.

French, J. W., Ekstrom, R. B., & Price, L. A. (1963). *Manual for kit of reference tests for cognitive factors*. Princeton, NJ: Educational Testing Service.

Guidance Testing Associates. (1962). *Prueba de Lectura, Nivel 5 - Advanzado Forma DEs.*

Guidance Testing Associates. (1962). *Test of Reading, Level 5 - Advanced Form CE.*

Joreskog, K. G., & Sorbom, D. (1981). *LISREL: Analyses of linear structural relationships by the method of maximum likelihood. User's guide, Version V.* Chicago: National Educational Resources.

Laosa, L. (1982). School, occupation, culture, and family: The impact of parental schooling on the parent-child relationship. *Journal of Educational Psychology, 74,* 791–827.

Laosa, L. (1984). Ethnic, socioeconomic, and home influences upon early performance measures of abilities. *Journal of Educational Psychology, 76,* 617–627.

Macnamara, J. (1967). The effects of instruction in a weaker language. *Journal of Social Issues, 23,* 121–135.

Macnamara, J., & Kellaghan, J. (1968). Reading in a second language. In M. Jenkinson (Ed.), *Improving reading throughout the world.* Newark, DE: International Reading Association.

Mestre, J. (1984). *On the relationship between the language of natural discourse and the language of logic: The interpretation of semantically complex sentences.* Amherst, MA: University of Massachusetts, Cognitive Processes Research Group. (ERIC Document Reproduction Service No. ED 268 819)

Manuel, H. T. (1963). *The preparation and evaluation of inter-language testing materials.* Austin, TX: University of Texas at Austin.

Oller, J. W., Jr. (1979). *Language tests at school.* London: Longman.

Chapter 15

Effects of Home Language and Primary Language on Mathematics Achievement:
A Model and Results for Secondary Analysis*

David E. Myers and Ann M. Milne

DRC

Washington, DC

The purpose of this study is to determine the effects of home and primary language on the mathematics achievement of language minority students, controlling for other background variables and investigating the importance of specific intervening mechanisms. Over the past two decades a large and growing body of research literature has dealt with the school-related performance of language minority students and the possible correlates of such performance. Although the research findings are diverse, in nearly every study one language minority group or another is shown to demonstrate school performance below that of native English speakers.

In the last decade much of the motivation for this research has centered around attempts to isolate effective teaching techniques for language minority populations, engendered in turn by federal policy positions in this area. The federal focus, both in regulations relating to the use of federal funds provided under Title VII of the Elementary and Secondary Education Act, and in monitoring of school programs for language minority populations by the U.S. Department of Education's Office for Civil Rights, has been on the provision of bilingual education. This focus, requiring the teaching of subject matter in two languages (either simultaneously or seriatim) has brought forth a number of studies designed to define, defend, or discredit bilingual education. Unfortunately, this has led to an

*This is a revised version of a much longer report prepared for Part C of the Math and Language-Minority Student Project, National Institute of Education. The opinions expressed here do not represent those of the Department of Education or DRC.

259

underemphasis on the possible correlates of the school performance of language minority students—both structural correlates such as family socioeconomic status and process correlates such as children's cognitive learning styles.

Although some studies of the correlates of school performance of language minority populations have attempted to fill this gap, many of them present one difficulty or another that make interpretation problematic. Greatest among these has been the lack of data bases with measures on major variables of probable relevance. This study was undertaken in an effort to overcome some of the problems presented by earlier studies. However, it is an unfortunate reality of data collection that although some data bases offer adequate measures of structural variables and others offer detailed clinical measures of process variables, we could find none that offer both. We therefore elected to investigate a model of primarily structural variables, based on the most recent and comprehensive of the large data bases—the base year measures for the High School and Beyond study.

Thus, much of the literature on language minority populations is beyond the scope of this study. For the reader interested in more detail on topics not covered here, we recommend the following bodies of literature:

- for evaluations of bilingual education and other pedagogic approaches, see the reviews by Dulay and Burt (1977, 1979); by Troike (1978); and the most recent by Baker and deKanter (1983);
- for studies of cognitive learning styles (including field dependence, self-concept, locus of control, etc.) evidenced by language-minority students, see studies by authors such as Hernandez (1973); Gerace and Mestre (1982); Kagan and Buriel (1977); and
- for clinical studies of the learning patterns of language-minority students, see studies by Gerace and Mestre (1982); Duran (1979); Macnamara (1966, 1967); and references listed by these authors.

LITERATURE REVIEW

As just noted, much of the literature reports that the school-related performance—attainment or achievement—of various language minority student populations is lower than that of native English speakers. This is most evident in situations in which the children are tested in English, the language of schooling, but has also been shown when children are tested in their own language (e.g., Anderson & Anderson, 1970) and/or on tests with essentially no verbal content (Anderson & Anderson, 1970; Morgan, 1957). Given that the largest language minority population in the United States is of Hispanic origin, the majority of studies have focused on this population. The results depend to some extent on whether the independent variable under study is truly language, and is the student's language at that, or whether the groups are simply classified on racial or ethnic lines. Results also depend on the extent to which other correlates are taken into account. Nevertheless, there is a substantial body of literature outlining decrements in the school-related performance of Hispanic students (e.g., Baral, 1979; Coleman et al., 1966; Evans & Anderson, 1973; Mayeske et al., 1973; Nielsen & Fernandez, 1981).

Two large and well-known reports that underline decrements in both school achievement and attainment in Hispanic populations are the NCES publication, *The Condition of Education for Hispanic Americans* (1980), and Carter and Segura (1979). The latter publication in particular stresses the degree of "fault" that should be attributed to the school rather than to the child or the child's culture, as do other studies by Felice (1978), by the U.S. Commission on Civil Rights (1973), and others. This is a legitimate concern, but is often not covered adequately in other studies, including the present one. It should be noted, however, that none of these studies includes measures of the child's primary language proficiency or proficiency in English, and although some studies demonstrate the lower socioeconomic status of Hispanics, none control for it.

There are few studies of language minority groups other than Hispanics and almost none that include groups that are monolingual in any non-English language. Earlier studies based on the High School and Beyond data (Myers & Milne, 1982; Stewart, Myers, & Milne, 1982) found that several non-English home language groups in the United States—Italian, German, Chinese—outperform native English speakers in a number of content areas (e.g., reading, mathematics, science), whereas those from a Spanish home background show decrements relative to native English speakers. In addition, Stewart et al. (1982) found that children who report both Spanish and English as their primary language have lower achievement than students who primarily speak Spanish.

Mayeske et al. (1973), in a reanalysis of the Coleman data, found that all ethnic groups tested ("Indian American, Mexican-American, Puerto-Rican, Negro, Oriental-American") had lower levels of achievement than "Whites." The State of California (California Assessment Program, 1980) has published mean reading and mathematics scores for all students in Grades 3 and 6 by ethnic group and language fluency. Although no significance tests are reported, reading means for fluent-English-speaking Chinese, Japanese, and Filipino students are above those for English-only students, whereas the means for fluent-English-speaking Spanish students are lower. Reading means for all non-fluent-English groups are below those of English-only students. In mathematics, however, the means for all groups surpass those for English-only students, with the exception of Spanish students, both those who are fluent-English speakers and those who are not. Matthews (1979), in a study of over 5,000 bilingual children in the Seattle school district, found their reading levels to be somewhat below the national norm, whereas their mathematics achievement levels were somewhat above. The bilingual groups studied, ranked from highest to lowest on percentage of students in the upper three stanines in reading, were Japanese, "Other,"

Chinese, Spanish, Korean, Philippine, Vietnamese, and Samoan. Although comparisons among each bilingual group and English-only students were not made, within each bilingual group, the students who were the most fluent in English scored higher in reading than did the less English-fluent students (fluency effects are not presented for mathematics).

These studies raise the issue of whether minority status per se confers achievement difficulties beyond those associated with language status, and whether those difficulties are greater for some ethnic groups than for others. One important way in which ethnic groups differ, which has not been controlled in any of the above-mentioned studies, is socioeconomic status. As Uhlenberg (1972) points out, Asian minorities in the United States, particularly Japanese-Americans, generally enjoy higher income and occupational status than do Hispanic groups.

Higher socioeconomic status, as well as the fact that their native English language is the majority language, has also been suggested as an explanation for the high achievement of English monolingual students taught in French immersion classes in Canada (e.g., Lambert & Tucker, 1972; Paulston, 1974). These students not only perform well in French, but show no decrement in English language proficiency or achievement. Within the United States, data are not yet available to test the effects of immersion classes on English language majority students.

The school-related performance of language minority students has most often been studied in subject areas, which because of their verbal content can be assumed to be closely related to language (e.g., reading), although areas such as mathematics that tend to have less verbal content have also been examined. Macnamara (1966) suggests that language minority students may be more hampered in logical/verbal mathematics achievement than in simple computation. Studies that compare the effects of language on both reading and mathematics achievement often find differences in the two areas. Evans and Anderson (1973) found Anglo-Americans to outperform Spanish-dominant but not English-dominant Mexican-Americans on an achievement test of language skills, but to significantly outperform both Mexican-American groups on mathematics achievement. However, De Avila (1980), Matthews (1979), and Rosenthal, Baker, and Ginsburg (1983) found mathematics achievement in elementary school students to be less related to language proficiency (De Avila, Matthews) or home dependence on Spanish (Rosenthal et al., 1983) than was reading achievement.

Within the range of studies of school-related performance of language minority students, measures of the student's actual proficiency in either English or the minority language are relatively rare. Many studies are based on ethnic or cultural group distinctions rather than on language usage or proficiency (e.g., Bender & Ruiz, 1974; Carter & Segura, 1979; NCES, 1980). Others use distinctions based on the language usually spoken in the student's home (e.g., Anderson & Anderson, 1970; Evans & Anderson, 1973; Mayeske et al., 1973; Rosenthal,

Baker, & Ginsburg, 1983; Rosenthal, Milne, Ellman, Ginsburg, & Baker, 1983; Veltman, 1980). If one wishes to examine the effects of language and not the aggregate effects of culture, the studies of greatest relevance are those that deal with the student's primary language—English or other—and the student's proficiency in that language.

The studies with "true" measures of proficiency range from those relying on self-reports or teacher reports of proficiency (California Assessment Program, 1980; Garcia, 1981; Matthews, 1979; Nielsen & Fernandez, 1981; So & Chan, 1982) to those with oral or written tests of language proficiency (De Avila, 1980; Morgan, 1957). As noted by Fishman and Terry (1968) self-reports of language proficiency can be quite accurate, and thus no attempt will be made here to distinguish between the classes of proficiency data offered in these studies.

As we move toward studies with better measures of students' primary language and their proficiency, we find that decrements are less severe for students who have greater reliance on, and proficiency in, English (De Avila, 1980; Morgan, 1957; Nielsen & Fernandez, 1981; So & Chan, 1982). This can also be demonstrated in studies such as Rosenthal, Baker, & Ginsburg, 1983 and Rosenthal, Milne, Ellman, Ginsburg and Baker, 1983, which include measures of degree of in-home reliance on English, but no measures of a child's primary language or English proficiency.

A student's proficiency in a non-English language has also been shown to be positively related to achievement (e.g., Nielsen & Fernandez, 1981) as well as to English proficiency (Nielsen & Fernandez, 1981; Rodriguez-Brown & Junker, 1980). This fact has led authors such as Duran (1979), Cummins (1980), and Oller (1980) to posit an underlying general proficiency factor related to reading achievement, regardless of the specific language. On the other hand, Nielsen and Lerner (1982) have suggested that proficiency in more than one language may create a handicap in the performance of other mental tasks because of the extra burden entailed in maintaining separate language systems. However, Oller (1980) posits a general proficiency factor related to achievement in all areas. Nevertheless, as noted above, there are indications that mathematics achievement is less dependent on English usage or proficiency than is reading achievement (California Assessment Program, 1980; De Avila, 1980; Matthews, 1979; Rosenthal, Baker, & Ginsburg, 1983; Rosenthal, Milne, Ellman, Ginsburg, & Baker, 1983).

Despite the findings that greater reliance on (proficiency in) English correlates positively with achievement, it does not always explain all of the residual variance in the difference between the achievement of language minority groups and native English speakers. Other correlates of achievement (i.e., those known to be related to achievement for all children), such as socioeconomic status (e.g., Coleman et al., 1966; Sewell & Hauser, 1976) thus become candidates for possible contribution to this residual variance.

Among the socioeconomic variables that are potentially relevant to the study of achievement in language-minority students are parents' education, and income. As Cummins (1979) points out, early research on the achievement of language-minority students generally overlooked the possible contribution of such variables, assuming all deficits were solely language-related. The ultimate in this reasoning is represented by the now-famous quotation from the UNESCO (1953) report that "it is axiomatic that the best medium for teaching a child is his mother tongue" (p. 11). After the great success of immersion programs for majority language children (mainly middle- or upper-class) in Canada, a number of studies (e.g., Bowen, 1977; Paulston, 1974) focused on sociocultural and socioeconomic factors almost to the exclusion of linguistic ones. This led to the opposite extreme interpretation as, for example, in Bowen's (1977) statement that "the choice of language of instruction in our schools is linguistically irrelevant" (p. 116).

Although a number of studies have examined both socioeconomic and linguistic factors, many of them (e.g., Anderson & Anderson, 1970; Evans & Anderson, 1973; Morgan, 1957; Paulston, 1974) simply stress the probable importance of both types of factors without attempting to assess their relative contributions. In recent years, five studies using large data bases (De Avila, 1980; Mayeske et al., 1973; Rosenthal, Milne, Ellman, Ginsburg, & Baker, 1983; So & Chan, 1982; Veltman, 1980) have attempted to weigh the relative importance of socioeconomic versus language factors, although none has done so entirely satisfactorily.

Mayeske et al. (1973), in a reanalysis of Coleman's Equality of Educational Opportunity data, tested language in the home against a broad cluster of other family background factors (including parental education, occupation, income, family stability, and family childrearing process). The authors concluded that a large relationship between language and achievement "is *not* present, and that the relationship that does exist can be explained largely by differences among students in Socio-Economic Status" (p. 80). The language measure used is home language rather than the child's primary language, and the socioeconomic indices incorporated a broader range of variables than is usual, including some family structure variables that other investigators (e.g., Mercy & Steelman, 1982; Milne, Myers, & Ginsburg, 1982) have found to be potent mediators of other socioeconimic variables.

Veltman (1980), in an analysis of the data from the Survey of Income and Education, concluded that nearly two thirds of the difference in attainment between Spanish and Anglo students was attributable to social class factors, even though the only measure of socioeconomic status available was parental educational attainment. Apart from the lack of data on language proficiency and other socioeconomic variables, another difficulty with the data base used is the availability only of attainment data (age in grade) as the dependent measure—a less sensitive variable than achievement test test scores.

Rosenthal, Milne, Ellman, Ginsburg, and Baker, (1983), using data from the

Sustaining Effects Study of Title I to measure the relative importance of socioeconomic status versus family (but not necessarily child) dependence on a non-English language, concluded that when socioeconomic status was controlled, there was little residual effect of language on reading and mathematics achievement level, and virtually none on school-year learning. A recent reanalysis of those data (Rosenthal, Baker, & Ginsburg, 1983) has found more residual variance among language groups after controlling for socioeconomic status. However, both studies suffer from a lack of language proficiency information.

De Avila (1980) has presented data on a large sample of Hispanic students including oral tests of both Spanish and English proficiency. Although the sample is not nationally representative, the author concludes that language proficiency is a more important determinant of reading and mathematics achievement than is socioeconomic status. Unfortunately, although the proficiency and achievement data are child level, the socioeconomic data used are at the school level.

More recently, So and Chan (1982) using the High School and Beyond data base, assessed the relative importance of language, ethnicity, and socioeconomic status in predicting achievement. They concluded that even after ethnicity and socioeconomic status were controlled, language contributed some 50% of the difference between Anglo and Hispanic reading achievement scores. Unfortunately, the construction of the language variable appears to tap language usage rather than proficiency, and the definition of ethnicity appears to have excluded large numbers of whites.

Despite the problems in each of these studies it appears that although socioeconomic status differences are clearly important in explaining achievement differences between Spanish speakers and native English speakers, a reasonable amount of unexplained variance remains. Whether this is all attributable to language, or partly to other unmeasured demographic, process, or other variables is not yet clear from the available research. Although much research has been conducted on the demographic correlates of achievement in general, little has been done within the context of language minority status.

Among the other correlates, gender is important in the current study because of the controversy in the literature over the difference between boys and girls in mathematics achievement, and the probable causes. Simple models relating gender and home or primary language to achievement (as well as interactions between the independent variables) have been previously presented by these investigators, along with summaries of the relevant literature (see Myers & Milne, 1982b; Stewart et al., 1982). To summarize briefly, our results on gender generally replicated those found in the literature (see Benhow & Stanley, 1980; Fennema, 1974; Fennema & Sherman, 1977, 1978; Sherman & Fennema, 1977). That is, girls generally outperformed, or were equal to, boys in both reading and mathematics in elementary grades, whereas high-school-aged boys generally enjoyed an advantage in mathematics. There were no interactions between home language and gender in effects on achievement among elementary students, but interactions were present among high school students.

There is also an extensive body of literature demonstrating that achievement, intelligence, achievement motivation, and so forth are negatively related to number of siblings (e.g., Mercy & Steelmann, 1982; Nuttal, Nuttal, Polit, & Hunter, 1976; Page & Grandon, 1979; Rosen, 1961; Svanum & Bringle, 1980; Zajonc, 1976). Although the results may differ with respect to the importance of birth order or of balance between younger and older siblings, all agree almost without exception on the negative effects of family size on achievement. A number of theories have been put forth to explain these results, most notably the confluence theory proposed by Zajonc (1976). (See Page & Grandon, 1979, for a critique of this theory as it applies to the prediction of individual differences.) Another theoretical perspective suggests that an increase in number of children is reflected in a dilution of parental resources (see, for example, Spaeth, 1976). No studies have been conducted on the simultaneous effects of language and number of siblings on achievement.

Another demographic variable of relevance to the current study is length of residence in the United States. As Hernandez-Chavez (1978) and others have noted, there is a tendency for immigrant populations in the United States to move from monolingual use of the original language to monolingual English in three generations. Among children in school, large proportions of those whose parents rely predominantly on another language, nevertheless rely heavily on English themselves, particularly in situations outside the home (see Nielsen and Fernandez, 1981). As discussed above, several studies have shown that increasing reliance on (proficiency in) English leads to increasing achievement. Thus, length of residence should be positively correlated with achievement as longer residence in the United States leads to increasing reliance on English. Baral (1979) in fact finds this to be the case for Mexican-American students. Rodriguez-Brown and Junker (1980), however, find no relationship, and Anderson and Johnson (1971) and Nielsen and Fernandez (1981) find negative relationships, all within Hispanic populations. Anderson and Johnson (1971) suggest that the inconsistencies of these findings may be a result of the degree of "language loyalty" in the specific communities in which Hispanics reside. Nielsen and Fernandez (1981) suggest that recent immigrants may outperform longer-term residents because they have had limited exposure to the "ghettoization" and ethnic discrimination that may characterize the milieu of earlier immigrants. However, it should be noted that of these studies, the one by Nielsen and Fernandez (1981) is the only one that is nationally representative.

Based upon the literature discussed above, we have developed a conceptual model incorporating a number of language, socioeconomic, and demographic variables as well as intervening variables. The objectives are twofold: first, to isolate the independent effects of language on achievement by controlling for other background variables, most critically measures of socioeconomic status;

second, to determine the intervening variables through which language affects achievement.

CONCEPTUAL MODEL

The conceptual model developed for this study is outlined generally in Fig. 15.1. We have examined two models in our analyses, each of which includes a unique measure of language minority status as the major exogenous variable. The first model uses a measure of the language usually spoken in the home, regardless of the student's current or primary language; the second uses a measure of the student's primary language. Although it has been argued that policy decisions regarding a student's eligibility for special language services should be based on a determination of the student's actual language usage and proficiency (e.g., Barnes, 1983; Birman & Ginsburg, 1983; Dulay & Burt, 1980), common practice at both the federal and local levels utilizes a first screening based on the presence of a non-English language in the home. Thus, the first model allows us to estimate the utility of partitioning students based on the presence of such a non-English home language. There are sufficient numbers of students in the sample who come from various non-English home language backgrounds to enable us to study these groups; however, very few high school seniors still have a non-English home language as a primary language. Thus, the primary language model can only contrast Spanish language speakers with English Primary students. In each model, the reference group for language is the sample of English

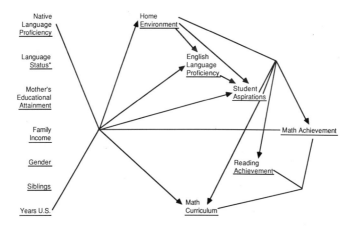

*Language Status refers to Home Language in Model 1 and Primary Language in Model 2.

FIG. 15.1.

Monolinguals—those students who reported English as their home language and that English is the only language they have ever spoken.

Along with the language measures, the other exogenous variables in the model include mother's educational attainment, family income, years of residence in the United States, proficiency in the non-English language, gender, and number of siblings. We have also included five endogenous or intermediate variables—educational home environment, English language proficiency, student aspirations, types of mathematics courses taken, and reading achievement. Each of the exogenous variables is hypothesized to have direct effects on each of the endogenous constructs in the model.

The placement of the endogenous variables in the model is somewhat problematic and subject to interpretation. For example, it may be argued that rather than aspirations influencing achievement, it may be that achievement influences aspirations. However, given that we are using cross-sectional data, it seems plausible that student aspirations, which are presumably based on students' past achievement, directly influence how well they are currently doing in school.[1]

The first of the endogenous variables, educational home environment, is measured by number of books in the home and presence of an encyclopedia. Many other investigators using the High School and Beyond data base use these two variables as part of a socioeconomic index (e.g., So & Chan, 1982; Nielson & Fernandez, 1981). However, Milne et al. (1982) and Spaeth (1976) suggest that these types of home environment variables act powerfully as intervening variables that mediate the effects of other socioeconomic variables. In the models, the assumption is that all of the exogenous variables have effects on, and operate through, the home environment. This general principle as it relates to the socioeconomic and demographic variables (i.e., mother's educational attainment, family income, years of residence, gender, and number of siblings) is confirmed by a number of authors (e.g., Spaeth, 1976; Williams, 1976). However, there is little in the literature that speaks directly to the probable effects of native language proficiency on home environment. The relevant literature (e.g., Cummins, 1979, 1980; Oller, 1980) does speak to a possible effect of native language proficiency on English language proficiency, and does not exclude the possibility that the effect may be indirect through home environment. Thus, we have elected to place English language proficiency after home environment in the model, and to estimate two potential effects of native language proficiency on it—one direct, and the other through home environment.

The available literature (e.g., Cummins, 1979, 1980; Nielsen & Fernandez, 1981; Oller, 1980) also posits a positive effect of native language proficiency on achievement. However, it is not clear whether the effect of native language proficiency on achievement is primarily a direct effect, or whether it operates via

---

[1]Ideally, longitudinal data would be used. However, at the time this study was completed, only the baseyear data from the High School and Beyond survey were available.

an increase in English proficiency. Within our model, the placement of native language proficiency as an exogenous variable and of English proficiency as an endogenous variable allows us to test both the direct and indirect effects on mathematics achievement. For the children of major interest here, the temporal ordering of the two proficiency variables is probably also correct, although it should be remembered that these are cross-sectional data—all measures were derived at one point in time.

The next intervening variable in the model, assumed to be affected by all prior variables, is the student's future aspirations, measured here by both educational plans and occupational aspirations. Although it would seem intuitive that students' English proficiency would condition their future educational plans, few studies of this potential link have been conducted. Portes, McLeod, and Parker (1978) in a study of Cuban refugees in 1973–74 found that their self-reported English proficiency did contribute positively to educational aspirations, but not to occupational aspirations; past educational attainment predicted both. The same study found that self-reported English proficiency of Mexican emigrants was a significant predictor of both educational and occupational aspirations.

Nielsen and Fernandez (1981) found that students with a better command of English have higher educational aspirations and that more frequent users of Spanish had lower educational aspirations. In addition, they found that high Spanish proficiency is also associated with high educational aspirations and that students whose families have resided longer in the United States have lower aspirations than recent immigrants. Bender and Ruiz (1974) found social class to be more important than ethnicity in determining aspirations, and that lower SES students, regardless of ethnicity, had both lower achievement and lower educational expectations. Evans and Anderson (1973) found that although Anglo- and Mexican-American students had similar occupational aspirations, Mexican-American students had lower educational aspirations. Interestingly, they found no relationship between socioeconomic status and aspirations for Mexican-Americans.

The final two intervening variables in the model are school-related variables hypothesized to be affected by all preceding variables, and to be correlated with one another. These variables are the type of mathematics courses taken by the students while in high school, and their reading achievement scores.

An earlier paper by these investigators (Myers & Milne, 1982a) tested the extent to which various language minority populations, and each gender, differentially elected to take various high school mathematics courses (as well as the interaction between gender and home language). There were a number of interactions present; nevertheless, it appeared that some home language groups were more likely to take advanced mathematics courses than others, and boys were more likely to do this than girls. To the extent that the content of such courses is relevant to the measure of mathematics achievement used here (and, as discussed below, that relationship is not entirely clear) it can be expected that students who have taken advanced mathematics courses would score higher on tests of mathematics achievement.

For reading achievement, the literature (e.g., Macnamara, 1966) suggests that achievement in this area will be a precursor to mathematics achievement to the extent that a mathematics test includes verbal, conceptual problems rather than straight computational problems. To some extent, the mathematics tests do have a verbal content, particularly the second and more difficult test used in the analyses reported here.

The final endogenous variables assumed to be affected by all preceding variables are two mathematics achievement tests. Both mathematics tests used in the High School and Beyond study were selected as outcome variables because they test different levels of mathematics achievement (Heyns & Hilton, 1982).

The data used in this analysis are from the 1980 High School and Beyond survey (HSB). The 1982 and 1984 follow-up data for High School and Beyond were not available when this research was commissioned. The HSB survey is a national longitudinal study of 28,000 high school seniors in the United States in 1980. (For the analyses reported here, we have included only the seniors because sophmores were not asked what types of mathematics courses they had taken.) Students for the survey were selected through a two-stage sampling design. Schools were first selected with probability proportional to their estimated enrollment; then within each school, 36 senior students were randomly selected. Because of the sample design, it is necessary to weight each of the student records used in the analyses. The weights provided with the HSB data base sum to national totals; we have normalized the weights to sum to the sample total. The variables used in the analyses are described in detail in the appendix.

METHODS

Each of the two models was estimated using a maximum likelihood procedure that allows for errors in equations (specification error) and errors in variables (measurement error). Thus, we are able to estimate a number of latent variables as well as the structural relationships among them. The models are specified and estimated in the LISREL framework (Joreskog & Sorbom, 1979, 1981).[2] The general structural model in matrix form is:

$$\eta = \beta\eta + \Gamma\xi + \zeta$$

and the measurement models in matrix form are

---

[2]The maximum likelihood estimation procedure assumes that the observed data are multivariate normally distributed. Provided with the available indicators and their measurement properties, this is clearly not a valid assumption and therefore, results of statistical significance must be interpreted with caution.

$$y = \lambda_y \eta + \epsilon$$

and

$$x = \lambda_x \xi + \delta$$

where β, Γ, λ_y, and λ_x are coefficient matrices, η and ξ are unobserved constructs, and y and x are observed variables. The structural models estimated here are hierarchical and nonrecursive in form. That is, there are no simultaneous relationships among latent variables and there are covariances between error terms in a number of equations.

The data from which each of the models is estimated are pair-wise covariance matrices. Pair-wise covariance matrices were used rather than matrices derived from list-wise deletion because of the large amount of missing data and the focus on a relatively small part of the over sample (i.e., language minority students). To confirm that use of the pair-wise data did not bias our results, we also estimated the models using data from the list-wise method and observed similar parameter estimates. (Univariate descriptive statistics and covariances for each of the variables discussed here can be found in Myers & Milne, 1983.)

RESULTS

Results for the structural model are discussed separately for the home language model and the primary language model. (The overall fit of each model to the data and the results for each measurement model are presented in Myers & Milne, 1983.)

Home Language Model

For the home language model, panel 3 of Table 15.1 shows summary measures of the model's ability to account for variation in the endogenous variables. The amount of explained variation ranges from a low of about .14 to a high of about .63 for home environment and the easier of the two mathematics tests, respectively.

Before discussing the effects of home language on mathematics achievement, we note that the direct (unmediated) effects of the other variables in the model such as students' gender and family income are in the direction that would be expected given the results of other studies in this area. The estimates of the direct effects are shown in panels 1 and 2 in Table 15.1.

Due to the large number of dimensions in the model, the discussion of the total, direct, and indirect effects is limited to those pertaining to the relationship between home language and each of the two mathematics tests and not the other

TABLE 15.1
Parameter Estimates Relating Exogenous and Endogenous Variables: Home Language Model

Endogenous Variables	Native Language Proficiency	Less than High School	High School Graduate	English Only	Spanish Only	Spanish/ English	Italian/ English	Chinese/ English	French/ English	German/ English	Other Language	Income	Gender	Siblings	Years US
(1) Home Environment	0.005	-0.085*	-0.033*	0.028	-0.104*	-0.032*	0.037	-0.021*	0.030	0.003	0.012	0.003*	0.011*	-0.001	0.005*
(2) English Language Proficiency	0.007*	-0.007*	0.000	-0.094*	-0.282*	-0.150*	-0.136	-0.363*	-0.143*	-0.099*	-0.147*	0.000	0.008*	-0.001	0.019*
(3) Student Aspirations	1.728*	-7.637*	-5.406*	-0.806	2.068	1.742	1.615	5.290	-0.076	-1.589	1.038	0.143*	5.405*	-0.280*	-0.259*
(4) Math Curriculum	0.001	-0.041*	-0.015*	-0.031	-0.035	-0.044*	0.025	0.189	0.030*	0.034	0.024	0.002*	-0.101*	-0.003*	-0.002
(5) READING	-1.341*	-1.023*	-0.370	0.226	1.699	0.840	2.736	6.829*	3.700*	4.345*	2.795*	0.051*	-1.571*	-0.267*	0.102
(6) MATH 1	-0.114	0.224	0.052	-1.293*	-1.460	-0.628	-0.245	1.393	0.615	1.265	0.339	0.041*	-1.052*	0.077	0.055
(7) MATH 2	0.613*	-0.147	-0.386*	-2.100*	-3.345*	-1.515*	-2.021	0.251*	-1.128	-0.048	-0.603	0.037*	-1.078*	0.055	0.010

Panel 2
ENDOGENOUS VARIABLES:

	Home Environment	English Language Proficiency	Student Aspirations	Math Curriculum	READ	MATH 1	MATH 2
(1) Home Environment	—						
(2) English Language Proficiency	0.053*	—					

	Home Environment	English Language Proficiency	Student Aspirations	Math Curriculum	READ	MATH 1	MATH 2
(3) Student Aspirations	16.322*	6.130*	—				
(4) Math Curriculum	0.138*	0.012	0.013*	—			
(5) READING	9.392*	5.986*	0.309*	—	—		
(6) MATH 1	1.622*	1.028	-0.018	24.101*	0.301*	—	
(7) MATH 2	-1.078	-0.752	0.001	20.920*	0.164*	—	—

Panel 3
ENDOGENOUS VARIABLES:

	Home Environment	English Language Proficiency	Student Aspirations	Math Curriculum	READ	MATH 1	MATH 2	
(1) Home Environment	0.015*							0.142
(2) English Language Proficiency	—	0.010*						0.319
(3) Student Aspirations	—	—	111.819*					0.226
(4) Math Curriculum	—	—	—	0.032*				0.493
(5) READING	—	—	—	0.507*	84.363*			0.213
(6) MATH 1	—	—	—	—	—	39.915*		0.628
(7) MATH 2	—	—	—	—	—	14.487*	63.707*	0.398

[a]Refers to the covariance matrix of errors in equations.
*Denotes significance at the .01 level.

endogenous variables in the model.[3] A decomposition of the total effect of each language group into direct and indirect effects allows us to assess the extent to which the intervening variables in the model account for the overall relationship between language minority status and mathematics achievement. For example, the decomposition allows us to determine if a negative (positive) total effect on mathematics achievement is primarily a function of a student's low (high) English language proficiency, low (high) level of mathematics background, or low (high) level of home environment. A discussion of the method for decomposing a variable's total effect (reduced form effect) into direct (unmediated), and indirect (mediated) effects is provided by Alwin and Hauser (1975), and Fox (1980).

The total, direct, and indirect effects for home language are shown in Table 15.2. All effects refer to a comparison between a language group and the English Monolinguals. This follows because home language refers to nine dummy variables with English Monolingual as the reference category. A total effect for Spanish as a home language can be interpreted, for example, as the difference between the total effect of the English Monolingual group (where the mean is by definition equal to zero) and those with a Spanish home language.

The total effect of being classified as being from a home where only English is spoken, but the student has had some exposure to a non-English language is -2.40 on the first mathematics test. This effect indicates that on the average, these students have scores that are about .25 standard deviations below the English Monolingual group. (For the entire sample of seniors, the standard deviation was set equal to 10.) Most of the total effect is accounted for by the direct (unmediated) effect of being classified as English Only (-1.29). To a lesser extent, the effect mediated by mathematics curriculum ($-.75$) contributes to the negative total effect. That is, being classified as English Only has a negative influence on mathematics curriculum, and this negative effect is transmitted via the positive effect of mathematics curriculum on mathematics achievement. For the second and more difficult of the two tests, nearly identical results are obtained—a significant total effect of -2.96 and a direct effect of -2.10 are estimated. In addition, mathematics curriculum appears to account for some of the negative total effect ($-.65$).

A comparison of Spanish Only students with the English Monolingual students provides a total effect of -4.04. That is, on the average, when controlling for the other background variables, Spanish Only students score about four tenths of a standard deviation below the English Monolingual students. Home environment and English language proficiency mediate much of the total effect, -1.50

---

[3]The 1% level of significance is used here rather than the conventional 5% level for two reasons. First, the sample is based on a stratified cluster design, and therefore, the effective sample size is smaller than the number of observations collected. Second, provided with the large number of significance tests reviewed here, it seems prudent to reduce the possibility of observing a large estimate by chance alone.

TABLE 15.2
Total, Direct and Indirect Effects of Home Language on Mathematics Achievement

Exogenous Variables	Total Effects	Direct Effects	Indirect Effects Via[a]					Endogenous Variables
			Home Environment	English Language Proficiency	Student Aspirations	Math Curriculum	Reading	
English Only	-2.40*	-1.29*	0.40	-0.52	-0.31	-0.75	0.07	MATH 1
	-2.96*	-2.10*	0.25	-0.23	-0.26	-0.65	0.04	MATH 2
Spanish Only	-4.04*	-1.46	-1.50	-1.55	0.80	-0.84	0.51	MATH 1
	-4.74*	-3.35*	-0.41	-0.70	0.67	-0.73	0.28	MATH 2
Spanish/English	-2.05*	-0.63	-0.46	-0.82	0.68	-1.06	0.25	MATH 1
	-2.38*	-1.52*	-0.28	-0.37	0.56	-0.92	0.14	MATH 2
Italian/English	1.59*	-0.25	0.53	-0.75	0.63	0.60	0.82	MATH 1
	-0.53	-2.02*	0.32	-0.34	0.52	0.52	0.45	MATH 2
Chinese/English	7.76*	1.39	-0.30	-2.00	2.05	4.56	2.05	MATH 1
	5.96*	0.25	-0.18	-0.89	1.71	3.95	1.12	MATH 2
French/English	2.07*	0.62	0.43	-0.79	-0.03	0.72	1.11	MATH 1
	-0.01	-1.13	0.26	-0.35	-0.02	0.63	0.61	MATH 2
German/English	2.27*	1.27*	0.04	-0.54	-0.62	0.82	1.30	MATH 1
	0.64	-0.05	0.03	-0.24	-0.51	0.71	0.71	MATH 2
Other Language	1.53*	0.34	0.17	-0.81	0.40	0.58	0.84	MATH 1
	0.44	-0.60	0.11	-0.36	0.34	0.50	0.45	MATH 2

*Denotes significance at the .01 level.

[a]Statistical significance not computed for indirect effects.

and −1.55, respectively. Thus, these indirect effects show that students who are classified as Spanish Only tend to have lower scores on the first mathematics achievement test because of lower levels of home environment and English language proficiency than English Monolingual students.

A different picture emerges for the results on the more difficult of the two mathematics tests. Again, a relatively large and significant negative total effect is estimated for the Spanish Only students (−4.74). However, a negative and large unmediated effect is also observed (−3.35). These results show that much of the total effect cannot be attributed to the intervening variables included in the model.

The total effect of a student residing in a home where both Spanish and English are spoken is −2.05. Mathematics curriculum appears to mediate much of the effect (−1.06). The remaining indirect effects are relatively small in comparison to the effect mediated by a student's mathematics curriculum. For the second test of mathematics achievement, a total effect of −2.38 and a significant direct effect of −1.52 are estimated. The relatively small mediated effects show that even after taking into account the other variables in the model, a direct and unmediated effect still persists. That is, there are factors not included in the model that contribute to the negative effect on mathematics achievement of being from a home in which Spanish and English are spoken, at least as measured by the second test.

The total effect on the first test of mathematics achievement for students from homes where both Italian and English are spoken in 1.59. In other words, after controlling for the other exogenous factors, Italian/English students score about .15 standard deviations above English Monolinguals. Home environment, student aspirations, mathematics curriculum, and reading achievement mediate positive effects, and English language proficiency mediates a negative effect. On the second mathematics test, a small, total effect is estimated (−.53). However, a negative unmediated effect is observed (−2.02). To some extent, the unmediated effect is moderated by the positive effects operating through home environment (.32), student aspirations (.52), mathematics curriculum (.52), and reading achievement (.45). English language proficiency mediates a negative effect (−.34).

For students from homes where Chinese and English are spoken a total effect of 7.76 is observed. This estimate shows that on the average, these students score about three quarters of a standard deviation above the English Monolinguals on the first mathematics achievement test. Much of this is a function of the effect mediated by mathematics curriculum (4.56), student aspirations (2.05), and reading achievement (2.05). For the second mathematics test a total effect of 5.96 is estimated. Mathematics curriculum accounts for a considerable fraction of the total effect (3.95). Reading achievement and student aspirations also mediate the total effect of residing in a home where both Chinese and English are spoken—1.12 and 1.71, respectively.

The total effect of speaking French and English in the home is 2.07. Part of this effect is positively mediated by reading achievement (1.11) and negatively mediated by English language proficiency (−.79). On the second test a small, insignificant negative total effect is estimated (−.01). None of the intervening variables appear to mediate much of the effect.

The final total effect on mathematics achievement to be decomposed is for those students who reside in homes where both German and English are spoken. The total effect for this group is 2.27. Much of this can be attributed to a significant unmediated effect of 1.27 and the path mediated by reading achievement (1.30). For the second mathematics test an insignificant total effect is estimated (.64). Most of this positive effect is produced by the effect of home language that is transmitted by mathematics curriculum (.71) and reading achievement (.71).

In general, the results for the home language model show that home language has relatively large, unmediated effects on mathematics achievement. That is, the factors included in the model do not transmit all of the effect of home language on mathematics achievement. However, we do observe that differences in home environment, English language proficiency, mathematics curriculum, and reading achievement do affect mathematics achievement to some extent. For example, those home language groups with higher attendance in advanced mathematics courses than the English Monolinguals had their mathematics achievement scores elevated to some extent. On the other hand, since each of the home language groups has, on the average, lower English language proficiency than the English Monolinguals, the students' scores are reduced by this factor. To some extent, the effects transmitted by English language proficiency and mathematics curriculum are moderated or enhanced by most home language groups having, on the average, higher student aspirations than the English Monolinguals; these aspirations in turn have positive effects on mathematics achievement.

Primary Language Model

The primary language model classifies the students based on the language they currently speak most often. Because most of these high school seniors generally rely on English, there are fewer non-English groups in the primary language model than in the home language model. Within the primary language model, we first discuss the direct effects of the exogenous (other than primary language) and endogenous variables.

Before discussing the actual estimates of the effects, we briefly describe the extent to which variation in each of the endogenous variables is accounted for by all prior variables in the model. We find that at one extreme, only about 14% of the variation in home environment is explained by the model. At the other extreme, we find that nearly 63% of the variation in test scores on the easier of the two math tests is accounted for by all prior variables in the model.

As in the home language model, we find that the estimate of the direct effects of the other variables (i.e., other than those measuring primary language) coincide with those in earlier studies. Estimates of the direct effects are shown in panels 1 and 2 of Table 15.3. The consistency between the home language and primary language models is to be expected given that the structures of the two models are similar and that we have merely substituted primary language for home language in the model.

Results of the decomposition of the total effects of a student's primary language on the two tests of mathematics achievement are shown in Table 15.4.

On the first mathematics test, the total effect of primarily speaking English but being familiar with another language, is 1.03. This shows that these students, in contrast to those who only speak English, score slightly higher on the easier of the two mathematics achievement tests. Much of this effect operates through students' reading achievement (.75) and English language proficiency (−.66). Home environment, student aspirations, and mathematics curriculum mediate only small fractions of the positive total effect. For the second mathematics test, a small negative total effect is estimated (−.34). To some extent, this effect is made up of a negative and significant unmediated effect (−1.73) as well as a mediated effect that is transmitted by English language proficiency (−.26). The negative impact of these two effects is moderated by a positive effect operating via reading achievement (.41).

The total effect on the first mathematics test of being classified as primarily speaking Spanish is −2.90. That is, these students score about one quarter of a standard deviation below English Monolinguals. Much of the negative total effect is accounted for by English language proficiency (−3.36) and home environment (−1.24). These negative indirect effects are moderated to some extent by the relatively high aspirations of these students (1.33). The unmediated effect of primarily speaking Spanish (−.33) is not statistically significant. On the second mathematics test, a negative total effect of −3.24 is estimated and an insignificant direct effect (−2.24) is observed. A relatively large indirect effect can be attributed to English language proficiency (−2.21).

The results for this group show that even after controlling for the family background variables, primarily speaking Spanish is significantly related to lower mathematics achievement. However, the estimates also show that the poor performance on the mathematics tests is a function of low English language proficiency and to some extent, home environment.

For Spanish/English students the total effect on the easier of the two mathematics tests is −7.90. Much of the total effect is accounted for by a large, negative and significant unmediated effect (−2.34). In addition, the scores are depressed by the negative effect transmitted via mathematics curriculum (−4.76). Additional negative effects are transmitted by reading achievement and English language profi-

TABLE 15.3
Parameter Estimates Relating Exogenous and Endogenous Variables: Primary Language Model

Panel 1 EXOGENOUS VARIABLES

Endogenous Variables	Native Language Proficiency	Less than High School	High School Graduate	English	Spanish	Spanish/English	Other Language	Income	Gender	Siblings	Years US
(1) Home Environment	-0.006	-0.087*	-0.033*	0.019*	-0.085*	0.032	0.034	0.003*	0.011*	-0.001	0.004*
(2) English Language Proficiency	0.013*	-0.009*	0.000	-0.140*	-0.469*	-0.233*	-0.348*	0.000	0.007*	0.000	0.016*
(3) Student Aspirations	1.843*	-7.579*	-5.413*	0.563	3.415	2.896	0.937	0.143*	5.409*	-0.281*	-0.252*
(4) Math Curriculum	-0.003	-0.041*	-0.014*	0.011	-0.031	-0.197*	-0.033	0.002*	-0.102*	-0.003*	-0.003*
(5) READING	-1.402*	-1.015*	-0.339	2.475*	0.983	-4.297*	-1.129	0.050*	-1.586*	-0.273*	0.045
(6) MATH 1	-0.233	0.212	0.064	0.180	-0.331	-2.342*	-0.361	0.041*	-1.048	0.074*	0.038
(7) MATH 2	0.556*	-0.171	-0.381*	-1.073*	-2.236	-3.669*	-0.872	0.036*	-1.072*	0.053	-0.004

Panel 2 ENDOGENOUS VARIABLES:

	Home Environment	English Language Proficiency	Student Aspirations	Math Curriculum	READ	MATH 1	MATH 2
(1) Home Environment	—						
(2) English	0.052*	—					

(Continued)

TABLE 15.3 (Continued)

Panel 1 EXOGENOUS VARIABLES

Endogenous Variables	Native Language Proficiency	Less than High School	High School Graduate	English	Spanish	Spanish/English	Other Language	Income	Gender	Siblings	Years US
Language Proficiency											
(3) Student Aspirations	16.201*	6.196*	—								
(4) Math Curriculum	0.144*	-0.008	0.013*	—							
(5) READING	9.548*	5.245*	0.309*	—	24.138*	0.302*					
(6) MATH 1	1.703*	0.911	-0.019	—	20.938*	0.164*					
(7) MATH 2	-1.016	-0.851	0.001	—	—	—					

Panel 3 ENDOGENOUS VARIABLES

Endogenous Variables	Home Environment	English Language Proficiency	Student Aspirations	Math Curriculum	READ	MATH 1	MATH 2	R^2
(1) Home Environment	0.015*							0.138
(2) English Language Proficiency		0.010*						0.352
(3) Student Aspirations		—	112.029*					0.225
(4) Math Curriculum		—	—	0.032*				0.492
(5) READING		—	—	0.509*	84.418*			0.213
(6) MATH 1		—	—	—	—	39.981*		0.628
(7) MATH 2		—	—	—	—	14.551*	63.780*	0.397

[a]Refers to the covariance matrix of errors in equations.

*Denotes significance at the .01 level.

TABLE 15.4
Total, Direct and Indirect Effects of Primary Language on Mathematics Achievement

Exogenous Variables	Total Effects	Direct Effects	Indirect Effects Via[a]					Endogenous Variables
			Home Environment	English Language Proficiency	Student Aspirations	Math Curriculum	Reading	
English Only	1.03*	0.18	0.27	-0.66	0.22	0.27	0.75	MATH 1
	-0.34	-1.73*	0.17	-0.26	0.18	0.23	0.41	MATH 2
Spanish Only	-2.90*	-0.33	-1.24	-3.36	1.33	-0.75	0.30	MATH 1
	-3.24*	-2.24	-0.76	-2.21	1.11	-0.65	0.16	MATH 2
Spanish/English	-7.90*	-2.34*	0.47	-1.67	1.12	-4.76	-1.30	MATH 1
	-7.71*	-3.67*	0.29	-1.10	0.94	-4.12	-0.70	MATH 2
Other Language	-2.28	-0.36	0.50	-2.49	0.36	-0.80	-0.34	MATH 1
	-1.79	-0.87	0.30	-1.64	0.30	-0.69	-0.19	MATH 2

*Denotes significance at the .01 level.

[a]Statistical significance not computed for indirect effects.

ciency, -1.30 and -1.67, respectively. The results for the second mathematics test are quite similar. A total effect of -7.71 and a negative direct effect of -3.67 are estimated. The negative total effect shows that these students on the average score about three quarters of a standard deviation below English Monolingual students when background variables are controlled. A large negative indirect effect that operates through mathematics curriculum is observed (-4.12). In addition, relatively small, negative effects are mediated by English language proficiency (-1.10) and reading achievement ($-.70$). It is interesting to note that although the model mediates the effect of primarily speaking Spanish on mathematics achievement, it does not do as well for those who report speaking both Spanish and English. In other words, even when the intervening variables and the background variables are controlled, the direct effects of speaking both Spanish and English are still statistically significant. This finding shows that there are other variables relevant to the Spanish/English speakers that are not included in the model.

DISCUSSION AND CONCLUSIONS

In the models analyzed for this study, we have demonstrated that there are significant differences between the mathematics achievement of various language status groups and that of monolingual English students. Controls for potentially important background variables—socioeconomic status, native language proficiency, gender, number of siblings, and length of residence in the United States—have not succeeded in reducing the total effects to insignificance in the majority of cases. We have been somewhat more successful in discovering the variables through which the effects of language status on achievement are mediated, although this varies by language status group and by the test used to measure mathematics achievement. The implications of these results vary according to which definition is used for "language status" (i.e., home and primary language).

The first model uses the language most often spoken in the home as the measure of language status. The results for the model show that after the background variables are controlled, the Italian/English, Chinese/English, French/English, German/English and "other" groups have significantly higher achievement scores on at least one mathematics test than do English Monolinguals, whereas English Only, Spanish Only and Spanish/English groups are significantly lower than English Monolinguals. The groups scoring significantly higher than English Monolinguals do so primarily on the first (longer) mathematics test, and the significant negative differences occur for both tests. As expected, the Spanish/English home language group is closer to the English Monolingual group than is the Spanish Only group—two tenths of a standard deviation as opposed to four tenths below.

In this model, we cannot interpret the residual language effects remaining after background variables are controlled as due to the effect of language per se on achievement. As the cross-tabular analysis of primary and home language showed, far more students from any of these home language groups rely on English than on the major non-English language spoken in the home. Thus, where significant effects of a non-English home language remain after the background variables are taken into account, they are the likely result of variables not included in the model. Possible candidates include measures of socioeconomic status not used here, such as father's occupation and mother's work status, and other cultural variables such as community structure and support.

The intervening variables in the home language model help to further explain the differences between the language groups and English Monolinguals, although not for all groups. Thus, the apparent large advantage in mathematics achievement of Chinese/English home language students is not a simple direct effect, but is positively affected by these students' exposure to advanced mathematics courses, by their reading as measured by the achievement test, and by their high educational and occupational aspirations. This is despite their relatively poor proficiency in English. The Spanish Only students, on the other hand, exhibit the same negative influence that operates via English proficiency, without the offsetting advantage of attendance in advanced mathematics courses. Also, although the effects of home language usage are transmitted positively through their aspirations and reading achievement, the effect is not as strong as for the Chinese/English (and some other) home language groups.

In summary, the intervening variables in the home language model are more effective at mediating the effects of home language for some groups than for others. That is, for five groups—English Only, Spanish Only, Spanish/English, Italian/English and German/English—there are significant direct (unmediated) effects of home language on at least one mathematics test; the intervening variables thus do little to explain the effects of home language on achievement. However, the particular intervening variables used have differential effectiveness in distinguishing among home language groups. Those groups outscoring the English Monolinguals on mathematics achievement are clearly helped by their exposure to advanced mathematics courses, and the lack of such exposure is clearly detrimental to the other groups. On the other hand, all groups without exception suffer a negative effect of their degree of English proficiency, relative to English Monolinguals (by definition, more proficient).

Within the primary language model, we can be more assured of dealing with the proximal effects of students' current language usage on mathematics achievement. Unfortunately, the number of students who primarily use languages other than Spanish or English are so few they must be grouped in an "other" category. Thus, the model primarily tests for the effects of various degrees of reliance on Spanish.

In this respect, the full model used here replicates the results of a simple model presented in an earlier paper (Stewart et al., 1982), where the only independent variables were primary language and gender. Despite the interactions between these two variables in their effect on mathematics achievement, it was apparent that students claiming both Spanish and English as a primary language had greater relative achievement deficits than students listing Spanish alone as a primary language. In the current analyses, the model is much less effective in explaining the Spanish/English deficit than in explaining the smaller Spanish Primary deficit. That is, the significant direct effect of Spanish/English primary language remains despite the inclusion of intervening variables and additional background variables. It appears that only the lack of attendance in advanced mathematics courses is an important intervening variable for Spanish/English students although this does not significantly reduce the direct effect. For the Spanish Primary students, the relatively smaller total deficit is in large part accounted for by their lack of English proficiency.

It is not clear why the relatively consistent finding in the literature that greater reliance on English should foster higher achievement is not upheld here. There are a number of potential explanations, none entirely satisfactory. First, students in the study were asked to indicate *the one* language they use primarily, and those who nevertheless insisted on identifying two (in this case, Spanish and English) may have some true confusion about their strongest language. It is also possible that the effort to develop proficiency in a second language causes proficiency loss in the first (Nielsen & Lerner, 1982), and the proficiency of the Spanish/English students here may be "comparably limited" in both languages (Dulay & Burt, 1980). In fact, the Spanish/English students do report lower proficiency in Spanish than the Spanish Primary students, but also report higher proficiency in English. It may be, of course, that self-reports of proficiency are less reliable for students who are not certain of their "primary" language.

If these students are truly bilingual, it may still cause difficulties in the (English-oriented) school situation. As Nielsen and Lerner (1982) note, the necessity of maintaining two language systems may cause handicaps in the performance of other mental tasks. This would appear to be supported by the fact that the Spanish/English Primary students not only have lower mathematics achievement than the Spanish Primary group, but for the former these effects are also mediated negatively through reading achievement, while being mediated positively through reading for the latter.

There are also differences in the backgrounds of these two groups of students, in particular the fact that the Spanish Primary students have lived in the United States approximately three years less than the Spanish/English students. Although this variable has been controlled for in the model, there may be additional correlates of immigrant status not accounted for here. Foremost among these is the effect of discrimination noted by others (e.g., Carter & Segura, 1979; Nielsen & Fernandez, 1981; Nielsen, Fernandez, & Peng, 1981).

In addition, there may be other variables not included in the model that account for the somewhat anomalous lower achievement of the Spanish/English students. Foremost among these missing variables may be those relating to schooling and teaching processes, community support variables, or other less tangible variables.

In summary, our analyses demonstrate significant differences in mathematics achievement between various language status groups and English Monolinguals, some lower and some higher. Although the importance of some intervening mechanisms is suggested—proficiency in English, and exposure to mathematics courses—the causal paths are not always clear, nor are the background correlates. The variables included in these models are only partly successful in clarifying the links between language status and mathematics achievement, and the inclusion of other variables in future research might be more elucidating.

APPENDIX
Description of Variables

Nine constructs in the model are measured with single indicators and include (1) student's home language; (2) student's primary language; (3) mother's educational attainment; (4) family income; (5) length of residence in the United States; (6) number of siblings; (7) gender; (8) reading achievement; and (9) mathematics achievement. Mother's educational attainment and family income are considered to provide socioeconomic characteristics of students, and demographic characteristics are provided through the inclusion of length of residence in the United States, number of siblings, and gender. In addition, five latent variables (constructs) are measured with two or more indicators: (1) English language proficiency; (2) native language proficiency; (3) educational home environment; (4) student aspirations; and (5) mathematics curriculum.

Student's home language was classified according to the following scheme:

1. English Monolingual;
2. English Only;
3. Spanish Only;
4. Spanish/English;
5. Italian/English;
6. Chinese/English;
7. French/English;
8. German/English; and
9. Other

Students were selected into the various home language categories on the basis of five questions asked in the HSB survey. The first question asked the first language spoken as a child. The second question asked what other language was spoken as a child—before starting school. The third question asked what language was usually spoken at present. The fourth question asked what language was primarily spoken in the home, and the final question asked whether another language was spoken in the home and if so, which language. Students were classified as being from English Monolingual homes if they indicated on all questions that only English was used. Students who indicated that they had some exposure to a non-English language but that only English was used in the home were classified as English Only. Those students who reported that only Spanish was spoken at home were classified at Spanish Only irrespective of the answers to the other questions. Students from homes where both Spanish and English were spoken were classified as Spanish/English, and so on. Because there were so few students from homes where *only* a non-English, non-Spanish language was spoken (e.g., only Chinese spoken at home) these students were pooled in the ''Other'' category. Table 15.A.1 presents the distribution of students on the home language variable.

The primary language of each student was classified according to the following scheme:

1. English Monolingual;
2. English Primary;
3. Spanish Primary;
4. Spanish/English Primary; and
5. Other Primary

Students' primary language was assessed on the basis of the same five questions used to classify home language. If the student indicated on all five ques-

TABLE 15.A.1
Frequency Distribution of Students by Home
Language

Home Language	Frequency
English monolingual	22,836
English Only	375
Spanish Only	531
Spanish/English	2,288
Italian/English	308
Chinese/English	54
French/English	289
German/English	348
Other	1,211

TABLE 15.A.2
Frequency Distribution of Students
by Primary Language

Primary Language	Frequency
English monolingual	22,836
English Primary	4,656
Spanish Primary	390
Spanish/English Primary	146
Other	212

tions that no language other than English was used, the student was classified as English Monolingual. If students indicated some current family use or personal childhood use of a non-English language, but indicated that they currently speak English most often, they were classified as English Primary. (Note that these children come from backgrounds with a variety of non-English languages.) If students indicated that the language they use most often is Spanish, regardless of home or early personal language usage, they were classified as Spanish Primary; students who reported that Spanish and English was their current primary language were classified as Spanish/English Primary. Because there were so few students with a non-English, non-Spanish primary language, the remaining students were classified as Other Primary.[4] In Table 15.A.2, the distribution of students according to primary language is shown.

A cross-tabulation of student's home and primary language verifies that these students confirm expectations based on the research literature. (Table not presented here, see Myers & Milne, 1983.) That is, among language-minority families there is clearly generational movement from reliance on a non-English language to reliance on English (Hernandez-Chavez, 1978). There is also a greater tendency for students from Spanish Only homes than from Spanish/English homes to currently rely only on Spanish. Among those students who currently use English primarily, there is a wide variety of non-English backgrounds. Finally, we note that there are 324 students whose families use only English and who themselves primarily speak English who nevertheless had some experience with a non-English language as children.

Mother's educational attainment is treated as a categorical variable, coded as less than high school graduation, high school graduation, and post high school education. Family income as reported by students is also incorporated into the

---

[4]Although the two language categories "other" home language and "other" primary language have been included for completeness, their relationship to mathematics achievement is not discussed because no substantive meaning can be attached to these categories.

model. Length of residence in the United States is obtained from a 4-point scale, based on a question in the HSB survey that asked students to indicate if they had lived in the United States 1 to 5 years, 6 to 10 years, over 10 years but not all of their life, and all or almost all of their life. These data were recorded as 3 years, 8 years, (10 + student's age)/2, and student's age, respectively.

Number of siblings was obtained from five questions that asked, for example, how many brother or sisters do you have who are 3 or more years older than you? Gender was coded as 0 for males and 1 for females.

The remaining constructs in the model are reading and mathematics achievement. The test used in the analyses to measure reading achievement is composed of eight items. This test corresponds to the "common items" administered to both sophomores and seniors as used in the 1980 HSB survey. Although reading tests with a greater number of items were available, the "common items" were used to maintain some comparability with previously published studies (Coleman, Hoffer, & Kilgore, 1982; Nielsen & Fernandez, 1981).

Two mathematics tests, one that reflects general mathematics ability and the other, a more advanced level of mathematics achievement, are used as the final outcome variables in the analyses. These tests were made up of 25 items and 8 items, respectively. The 8-item test has been suggested as being the more difficult of the two by Heyns and Hilton (1982). In all analyses described in this chapter, we treat the two tests separately in an effort to determine if the students are differentiated by the two mathematics tests. Further discussion of these and other tests administered in the 1980 High School and Beyond survey is provided by Heyns and Hilton (1982).

The latent variables measuring language proficiency, both exogenous and endogenous, were each measured by multiple indicators. English and native language proficiency are each measured by four indicators, based on questions that asked how well students understand, speak, read, and write in English and a native language. English Monolingual students were assigned a score corresponding to no ability, on each of the four indicators for native language proficiency. These same students are also assumed to have the highest possible scores on the indicators for English language proficiency. This procedure was used because the proficiency questions were not asked of the English Monolingual students. A similar procedure was followed by Nielsen and Fernandez (1981).

Educational home environment was determined by student self-reports of whether there were 50 or more books in their home, and if an encyclopedia set was present in their home. Student aspirations refers to whether students plan to attend college and what type of occupation they expect to be employed in when they are 30 years of age. College plans were coded as 1 for those who do not plan to attend, 2 for those who do not know, and 3 for those who do plan to attend college. Occupational aspirations was originally a 12-point, categorical variable;

however, we transformed the responses into Duncan SEI status scores to obtain a continuous variable. Students who aspired to be in the military or planned to be housewives were excluded from the analyses because of the ambiguity of their status.

Mathematics curriculum is measured by five indicators: (1) whether a student has taken Algebra 1; (2) whether a student has taken Algebra 2; (3) whether a student has taken Geometry; (4) whether a student has taken Trigonometry; and (5) whether a student has taken Calculus.

REFERENCES

Alwin, D. F., & Hauser, R. M. (1975). The decomposition of effects in path analysis. *American Sociological Review, 40,* 37–47.

Anderson, G. V., & Anderson, H. T. (1970). *Comparison of performance on a mental ability test of English speaking and Spanish speaking children in grades two and three.* Washington, DC: National Institute of Education. (ERIC Document Reproduction Service No. ED 090 335)

Anderson, J. G., & Johnson, W. H. (1971). Stability and change among three generations of Mexican-Americans: Factors affecting achievement. *American Educational Research Journal, 8* (2), 285–309.

Baker, K. A., & de Kanter, A. A. (Eds.). (1983). *Bilingual education.* Lexington, MA: D.C. Heath.

Baral, D. P. (1979). Academic achievement of recent immigrants from Mexico. *NABE Journal, 3,* 3, 1–13.

Barnes, R. E. (1983). The size of the eligible language-minority population. In K. A. Baker & A. A. de Kanter (Eds.), *Bilingual education.* Lexington, MA: D.C. Heath.

Bender, P. S., & Ruiz, R. A. (1974). Race and class as differential determinants of underachievement and underaspiration among Mexican-Americans and Anglos. *Journal of Educational Research, 68,* 2, 51–55.

Benbow, C. P., & Stanley, J. C. (1980). Sex differences in mathematical ability: Fact or artifact? *Science, 210,* 1262–1264.

Bentler, P. M., & Bonett, D. G. (1980). Significance tests and goodness of fit in the analysis of covariance structures. *Psychological Bulletin, 88*(3), 588–606.

Birman, B. F., & Ginsburg, A. L. (1983). Introduction: Addressing the needs of language-minority children. In K. A. Baker & A. A. de Kanter (Eds.), *Bilingual education.* Lexington, MA: D.C. Heath.

Bowen, J. D. (1977). Linguistic perspectives on bilingual education. In B. Spolsky & R. Cooper (Eds.), *Frontiers of bilingual education.* Rowley, MA: Newbury House.

California Assessment Program. (1980). *Student achievement in California schools 1979–80 Annual Report.* Sacramento: California State Department of Education.

Carter, T. P., & Segura, R. D. (1979). *Mexican-Americans in school: A decade of change.* New York: College Entrance Examination Board.

Coleman, J. S., Campbell, E. Q., Hobson, D. J., McPartland, J., Mood, A. M., Weinfeld, F. D., & York, R. L. (1966). *Equality of educational opportunity.* Washington, DC: U.S. Government Printing Office.

Coleman, J. S., Hoffer, T., & Kilgore, S. (1982). *High school achievement: Public, Catholic, and private schools compared.* New York: Basic Books.

Cummins, J. (1979). Linguistic interdependence and the educational development of bilingual children. *Review of Educational Research, 49*, 2, 222–251.

Cummins, J. (1980). The construct of language proficiency in bilingual education. In J. E. Alatis (Ed.), *Current issues in bilingual education.* Washington, DC: Georgetown University Press.

De Avila, E. A. (1980). *Relative language proficiency types: A comparison of prevalence, achievement level and socio-economic status.* Report submitted to the RAND Corporation, February 2, 1980.

Dulay, H., & Burt, M. (1977). Remarks on creativity in language acquisition. In M. Burt, H. Dulay, & M. Finocchiaro (Eds.), *Viewpoints on English as a second language* (pp. 95–126). New York: Regents.

Dulay, H., & Burt, M. (1979). Research priorities in bilingual education. *Education Evaluation and Policy Analysis, 1*(3), 39–53.

Dulay, H., & Burt, M. (1980). The relative proficiency of limited English proficient students. In J. E. Alatis (Ed.), *Current issues in bilingual education.* Washington, DC: Georgetown University Press.

Duran, R. P. (1979). *Logical reasoning skills of Puerto Rican bilinguals.* Princeton: Educational Testing Service.

Evans, F. B., & Anderson, J. G. (1973). The psychocultural origins of achievement and achievement motivation: The Mexican-American family. *Sociology of Education, 46*, 396–416.

Felice, L. G. (1978, March). *Mexican-American achievement performance: Linking the effects of school and family expectations to benefit the bilingual child.* Paper presented at the Annual Meeting of the American Educational Research Association, Toronto, Canada. (ERIC Document Reproduction Service No. ED 151 125)

Fennema, E. H. (1974). Mathematics learning and the sexes: A review. *Journal for Research in Mathematics Education,* May, 126–139.

Fennema, E. H., & Sherman, J. A. (1977). Sex-related differences in mathematics achievement, spatial visualization and affective factors. *American Educational Research Journal, 14*(1), 51–71.

Fennema, E. H., & Sherman, J. A. (1978). Sex-related differences in mathematics achievement and related factors: A further study. *Journal for Research in Mathematics Education,* May, 189–203.

Fishman, J. A., & Terry, C. (1968). The contrastive validity of Census data on bilingualism in a Puerto Rican neighborhood. In J. A. Fishman, R. L. Cooper, & R. Ma (Eds.), *Bilingualism in the barrio. Final report.* New York: Yeshiva University (sponsored by the Bureau of Research, Office of Education, U.S. Department of Health, Education, and Welfare).

Fox, J. (1980). Effect analysis in structural equation models. *Sociological Methodology and Research, 9*, 3–28.

Garcia, H. D. C. (1981). *Bilingualism, confidence, and college achievement* (Report no. 318). Baltimore: Center for Social Organization of Schools, The Johns Hopkins University.

Gerace, W. J., & Mestre, J. P. (1982). *A study of the cognitive development of Hispanic adolescents learning algebra using clinical interview techniques.* Amherst, MA: University of Massachusetts.

Hernandez, N. G. (1973). Variables affecting achievement of middle school Mexican-American students. *Review of Education Research, 43*, 1, 1–39.

Hernandez-Chavez, E. (1978). Language maintenance, bilingual education, and philosophies of bilingualism in the United States, In J. Alatis (Ed.), *International dimensions of bilingual education.* Washington, DC: Georgetown University Press.

Heyns, B., & Hilton, T. L. (1982). The cognitive tests for High School and Beyond: An assessment. *Sociology of Education, 55*, 89–102.

Joreskog, K. G., & Sorbom, D. (1979). *Advances in factor analysis and structural equation models*. Cambridge, MA: Abt Books.

Joreskog, K. G., & Sorbom, D. (1981). *LISREL analysis of linear structural relationships by the method of maximum likelihood: User's guide*. Uppsala, Sweden: University of Uppsala.

Kagan, S., & Buriel, R. (1977). Field dependence-independence and Mexican-American culture and education. In J. L. Martinez (Ed.), *Chicano psychology*, New York: Academic Press.

Lambert, W. E., & Tucker, G. R. (1972). *Bilingual education of children: The St. Lambert experiment*. Rowley, MA: Newbury House.

Macnamara, J. (1966). *Bilingualism in primary education*. Edinburgh: Edinburgh University Press.

Macnamara, J. (1967). The effects of instruction in a weaker language. *Journal of Social Issues, 23*(2), 121–135.

Matthews, T. (1979). *An investigation of the effects of background characteristics and special language service on the reading achievement and English fluency of bilingual students* Report No. 79-19). Seattle: Seattle Public Schools.

Mayeske, G. W., Okada, T., Beaton, A. E., Jr., Cohen, W. M., Wisler, C. E., & Mood, A. M. (1973). *A study of the achievement of our nation's students*. Washington, DC: U.S. Government Printing Office.

Mercy, J. A., & Steelman, L. C. (1982). Familial influence on the intellectual attainment of children. *American Sociological Review, 47*, 532–542.

Milne, A. M., Myers, D. E., & Ginsburg, A. L. (1982, March) *Single parents, working mothers, socioeconomic status, and children's achievement*. Paper presented at the Annual Meeting of the American Educational Research Association, New York.

Morgan, E. R. (1957). *Bilingualism and non-verbal intelligence: A study of test results, Pamphlet No. 4*. Aberystwyth, Wales: Wales University. (ERIC Document Reproduction Service No. ED 117 959).

Myers, D. E., & Milne, A. M. (1982a). *Mathematics opportunity and exposure: Language minority status and gender*. Prepared for Part C of the Math and Language Minority Students Project, National Institute of Education.

Myers, D. E., & Milne, A. M. (1982b). *Achievement by gender and home language for elementary and secondary school children: A preliminary analysis*. Prepared for Part C of the Math and Language Minority Students Project, National Institute of Education.

Myers, D. E., & Milne, A. M. (1983). *Mathematics achievement of language-minority students*. Prepared for Part C of the Math and Language Minority Students Project, National Institute of Education.

National Center for Education Statistics. (1980). *The condition of education for Hispanic Americans*. Washington, DC: U.S. Government Printing Office.

National Center for Education Statistics. (1980). High school and beyond survey. Washington, DC: U.S. Government Printing Office.

Nielsen, F., & Fernandez, R. M. (1981). *Achievement of Hispanic students in American high schools: Background characteristics and achievement*. Washington, DC: National Center for Education Statistics.

Nielsen, F., Fernandez, R. M., & Peng, S. S. (1981). *Achievement of Hispanic students in American high schools: Background characteristics and achievement*. Washington, DC: National Center for Education Statistics.

Nielsen, F., & Lerner, S. J. (1982). *Language skills and school achievement of bilingual Hispanics*. Prepared under research grant from the Spencer Foundation. University of North Carolina at Chapel Hill.

Nuttall, E. V., Nuttall, R. L., Polit, D., & Hunter, J. B. (1976). The effects of family size, birth order, sibling separation and crowding on the academic achievement of boys and girls. *American Educational Research Journal, 13*(3), 217–223.

Oller, J. W., Jr. (1980). A language factor deeper than speech: More data and theory for bilingual assessment. In J. E. Alatis (Ed.), *Current issues in bilingual education*. Georgetown University Roundtable in Languages and Linguistics (pp. 14–30) Washington, DC: Georgetown University Press.

Page, E. B., & Grandon, G. M. (1979). Family configuration and mental ability: Two theories contrasted with U.S. data. *American Educational Research Journal, 16,* 3, 257–272.

Paulston, C. P. (1974). Implications of language learning theory for language planning: Concerns in bilingual education. *Bilingual Education Series:* Papers in Applied Linguistics *1*. Arlington, VA: Center for Applied Linguistics.

Portes, A., McLeod, S., & Parker, R. (1978). Immigrant aspirations. *Sociology of Education, 51,* 241–260.

Rodriguez-Brown, F. V., & Junker, L. K. (1980). *The relationship of student and home variables to language proficiency and reading achievement of bilingual children*. Washington, DC: National Institute of Education. (ERIC Document Reproduction Service No. ED 193 942)

Rosen, B. C. (1961). Family structure and achievement motivation. *American Sociological Review, 26,* 574–585.

Rosenthal, A. S., Baker, K., & Ginsburg, A. (1983). *The effect of language background on achievement level and learning among elementary school students*. Unpublished manuscript.

Rosenthal, A. S., Milne, A. M., Ellman, F. M., Ginsburg, A. L., & Baker, K. A. (1983). A comparison of the effects of language background and socioeconomic status on achievement among elementary-school students. In K. A. Baker & A. A. de Kanter (Eds.), *Bilingual education*. Lexington, MA: D.C. Heath.

Sewell, W. H., & Hauser, R. M. (1976). Causes and consequences of higher education: Models of the status attainment process. In W. H. Sewell, R. M. Hauser, D. L. Featherman (Eds.), *Schooling and achievement in American society*. New York: Academic Press.

Sherman, J. A., & Fennema, E. H. (1977). The study of mathematics by high school girls and boys: Related variables. American Educational Research Journal, 14, 2, 159–168.

So, A. Y., & Chan, K. S. (1982). *What matters? A study of the relative impact of language background and socioeconomic status on reading achievement*. Los Alamitos, CA: National Center for Bilingual Research.

Spaeth, J. L. (1976). Cognitive complexity: A dimension underlying the socioeconomic achievement process. In W. H. Sewell, R. M. Hauser, & D. L. Featherman (Eds.), *Schooling and achievement in American society*. New York: Academic Press.

Stewart, S. K., Myers, D. E., & Milne, A. M. (1982). *Achievement by gender and primary language for secondary school students: A preliminary analysis*. Prepared for Part C of the Math and Language Minority Students Project, National Institute of Education.

Svanum, S., & Bringle, R. G. (1980). Evaluation of confluence model variables on IQ and achievement test scores in a sample of 6- to 11-year-old children. *Journal of Educational Psychology, 72*(4), 427–436.

Troike, R. (1978). Research evidence for the effectiveness of bilingual education. *NABE Journal, 3,* 13–24.

Uhlenberg, P. (1972). Demographic correlates of group achievement: Contrasting patterns of Mexican-Americans and Japanese-Americans. *Demography, 9*(1), 119–128.

United Nations Educational, Scientific and Cultural Organization. (1953). The use of vernacular languages in education. *Monographs on Fundamental Education*.

U.S. Commission on Civil Rights. (1971). *Mexican-American education study*. Washington, DC: U.S. Government Printing Office.

U.S. Commission on Civil Rights. (1973). *Mexican-American education study*. Washington, DC: U.S. Government Printing Office.

U.S. Department of Education, National Center for Education Statistics. (1980). High School and Beyond Survey,

Veltman, C. J. (1976). *Relative educational attainments of Hispanic American children.* Report prepared for the Aspira Center for Educational Equity.

Veltman, C. J. (1980, October). *Relative educational attainment of Hispanic-American children, 1976.* Paper presented at the Aspira Hispanic Forum for Responsive Educational Policy, Washington, DC.

Williams, T. (1976). Abilities and environments. In W. H. Sewell, R. M. Hauser, & D. L. Featherman (Eds.), *Schooling and achievement in American society* (pp. 61–101) New York: Academic Press.

Zajonc, R. B. (1976). Family configuration and intelligence. *Science, 192,* 227–236.

A Final Note. . .

We have chosen to conclude this volume with an essay that puts the "person" back into our attempts to capture the critical research variables. One set of variables that we have termed *opportunity to learn* included the affordances and appurtenances of well-trained teachers and well-equipped classrooms. Research, however, rarely captures the valences of events or experiences that affectively charge learning.

In a computer and electronics age we are confronted, when we consider accounts like the following, by the uncomfortable consideration of what constitutes "opportunity for learning," and where the "lost opportunities" loom. Does membership necessarily have all the privileges? In the Epilogue, Joseph Suina's account of a child's attempt to make sense of two worlds forces researchers and educators alike to consider what they would categorize as the "important" variables in our study of the language–cognition–learning interaction.

—The Editors

Epilogue: *And Then I Went To School*

Joseph H. Suina

University of New Mexico at Albuquerque

I lived with my grandmother from the ages of 5 through 9. It was the early 1950s when electricity had not yet invaded the homes of the Cochiti Indians. The village day school and health clinic were first to have it and to the unsuspecting Cochiti's this was the approach of a new era in their uncomplicated lives.

Transportation was simple then. Two good horses and a sturdy wagon met most needs of a villager. Only five or six individuals possessed an automobile in the Pueblo of 300. A flatbed truck fixed with wooden rails and a canvas top made a regular Saturday trip to Santa Fe. It was always loaded beyond capacity with Cochitis taking their wares to town for a few staples. With an escort of a dozen barking dogs, the straining truck made a noisy exit, northbound from the village.

During those years, Grandmother and I lived beside the plaza in a one-room house. It consisted of a traditional fireplace, a makeshift cabinet for our few tin cups and dishes, and a wooden crate that held our two buckets of all-purpose water. At the far end of the room were two rolls of bedding we used as comfortable sitting "couches." Consisting of thick quilts, sheepskin, and assorted blankets, these bed rolls were undone each night. A wooden pole the length of one side of the room was suspended about 10 inches from the ceiling beams. A modest collection of colorful shawls, blankets, and sashes draped over the pole making this part of the room most interesting. In one corner was a bulky metal trunk for our ceremonial wear and few valuables. A dresser, which was traded for some of my grandmother's well-known pottery, held the few articles of clothing we owned and the "goody bag." Grandmother always had a flour sack filled with candy, store bought cookies, and Fig Newtons. These were saturated with a sharp odor of moth balls. Nevertheless, they made a fine snack with coffee before we turned in for the night. Tucked securely in my blankets, I listened to one of her stories or accounts of how it was when she was a little girl. These accounts seemed so old fashioned compared to the way we lived. Sometimes she softly sang a song from a ceremony. In this way I fell asleep each night.

Earlier in the evening we would make our way to a relative's house if someone had not already come to visit us. I would play with the children while the adults caught up on all the latest. Ten-cent comic books were finding their way into the Pueblo homes. For us children, these were the first link to the world beyond the Pueblo. We enjoyed looking at them and role playing as one of the heroes rounding up the villains. Everyone preferred being a cowboy rather than an Indian because cowboys were always victorious. Sometimes, stories were related to both children and adults. These get-togethers were highlighted by refreshments of coffee and sweet bread or fruit pies baked in the outdoor oven. Winter months would most likely include roasted pinon nuts and dried deer meat for all to share. These evening gatherings and sense of closeness diminished as the radios and televisions increased over the following years. It was never to be the same again.

The winter months are among my fondest recollections. A warm fire crackled and danced brightly in the fireplace and the aroma of delicious stew filled our one-room house. To me the house was just right. The thick adobe walls wrapped around the two of us protectingly during the long freezing nights. Grandmother's affection completed the warmth and security I will always remember.

Being the only child at Grandmother's, I had lots of attention and plenty of reasons to feel good about myself. As a pre-schooler, I already had the chores of chopping firewood and hauling in fresh water each day. After "heavy work," I would run to her and flex what I was certain were my gigantic biceps. Grandmother would state that at the rate I was going I would soon attain the status of a man like the adult males in the village. Her shower of praises made me feel like the Indian Superman of all times. At age 5, I suppose I was as close to that concept of myself as anyone.

In spite of her many years, grandmother was still active in the village ceremonial setting. She was a member of an important women's society and attended all the functions taking me along to many of them. I would wear one of my colorful shirts she handmade for just such occasions. Grandmother taught me the appropriate behavior at these events. Through modeling she taught me to pray properly. Barefooted, I would greet the sun each morning with a handful of cornmeal. At night I would look to the stars in wonderment and let a prayer slip through my lips. I learned to appreciate cooperation in nature and my fellowmen early in life. About food and material things, grandmother would say, "There is enough for everyone to share and it all comes from above, my child." I felt very much a part of the world and our way of life. I knew I had a place in it and I felt good about me.

At age 6, like the rest of the Cochiti 6-year-olds that year, I had to begin my schooling. It was a new and bewildering experience. One I will not forget. The strange surroundings, new concepts about time and expectations, and a foreign tongue were overwhelming to us beginners. It took some effort to return the second day and many times thereafter.

To begin with, unlike my grandmother, the teacher did not have pretty brown skin and a colorful dress. She was not plump and friendly. Her clothes were one color and drab. Her pale and skinny form made me worry that she was very ill. I thought that explained why she did not have time just for me and the disappointed looks and orders she seemed to always direct my way. I didn't think she was so smart because she couldn't understand my language. "Surely that was why we had to leave our 'Indian' at home." But then I did not feel so bright either. All I could say in her language was "yes teacher," "my name is Joseph Henry," and "when is lunch time." The teacher's odor took some getting used to also. In fact, many times it made me sick right before lunch. Later, I learned from the girls that this odor was something she wore called perfume.

The classroom too had its odd characteristics. It was terribly huge and smelled of medicine like the village clinic I feared so much. The walls and ceiling were artificial and uncaring. They were too far from me and I felt naked. The fluorescent light tubes were eerie and blinked suspiciously above me. This was quite a contrast to the fire and sunlight that my eyes were accustomed to. I thought maybe the lighting did not seem right because it was man-made, and it was not natural. Our confinement to rows of desks was another unnatural demand from our active little bodies. We had to sit at these hard things for what seemed like forever before relief (recess) came midway through the morning and afternoon. Running carefree in the village and fields was but a sweet memory of days gone by. We all went home for lunch because we lived within walking distance of the school. It took coaxing and sometimes bribing to get me to return and complete the remainder of the school day.

School was a painful experience during those early years. The English language and the new set of values caused me much anxiety and embarrassment. I could not comprehend everything that was happening but yet I could understand very well when I messed up or was not doing so well. The negative aspect was communicated too effectively and I became unsure of myself more and more. How I wished I could understand other things just as well in school.

The value conflict was not only in school performance but in other areas of my life as well. For example, many of us students had a problem with head lice due to "the lack of sanitary conditions in our homes." Consequently, we received a severe shampooing that was rough on both the scalp and the ego. Cleanliness was crucial and a washing of this type indicated to the class how filthy a home setting we came from. I recall that after one such treatment I was humiliated before my peers with a statement that I had "She'na" (lice) so tough that I must have been born with them. Needless to say, my Super Indian self-image was no longer intact.

My language, too, was questionable from the beginning of my school career. "Leave your Indian (language) at home" was like a trademark of school. Speaking it accidentally or otherwise was a sure reprimand in the form of a dirty look or a whack with a ruler. This punishment was for speaking the language of my people which meant so much to me. It was the language of my grandmother and I spoke it well. With it, I sang beautiful songs and prayed from my heart. At that young and tender age, comprehending why I had to part with it was most difficult for me. And yet at home I was encouraged to attend school so that I might have a better life in the future. I knew I had a good village life already but this was communicated less and less each day I was in school.

As the weeks turned to months, I learned English more and more. It would appear comprehension would be easier. It got easier to understand all right. I understood that everything I had and was a part of was not nearly as good as the white man's. School was determined to undo me in everything from my sheep-skin bedding to the dances and ceremonies that I learned to believe in and cherish. One day I fell asleep in class after a sacred all-night ceremony. I was startled to awakening by a sharp jerk on my ear and informed coldly, "That ought to teach you not to attend 'those things' again." Later, all alone I cried. I could not understand why or what I was caught up in. I was receiving two very different messages; both intending to be for my welfare.

Life-style values were dictated in various ways. The Dick and Jane reading series in the primary grades presented me with pictures of a home with a pitched roof, straight walls, and sidewalks. I could not identify with these from my Pueblo world. However, it was clear I did not have these things and what I did have did not measure up. At night, long after grandmother went to sleep, I would lay awake staring at our crooked adobe walls casting uneven shadows from the light of the fireplace. The walls were no longer just right for me. My life was no longer just right. I was ashamed of being who I was and I wanted to change right then and there. Somehow it became so important to have straight walls, clean hair and teeth, and a spotted dog to chase after. I even became critical and hateful toward my bony, fleabag of a dog. I loved the familiar and cozy surroundings of grandmother's house but now I imagined it could be a heck of a lot better if only I had a white man's house with a bed, a nice couch, and a clock. In school books, all the child characters ever did was run around chasing their dog or a kite. They were always happy. As for me, all I seemed to do at home was go back and forth with buckets of water and cut up sticks for a lousy fire. "Didn't the teacher say that drinking coffee would stunt my growth?" "Why couldn't I have nice tall glasses of milk so I could have strong bones and white teeth like those kids in the books?" "Did my grandmother really care about my well-being?"

I had to leave my beloved village of Cochiti for my education beyond Grade 6. I left to attend a Bureau of Indian Affairs boarding school 30 miles from home. Shined shoes and pressed shirt and pants were the order of the day. I managed to adjust to this just as I had to most of the things the school shoved at me or took away from me. Adjusting to leaving home and the village was tough indeed. It seemed the older I got, the further away I became from the ways I was so much a part of. Because my parents did not own an automobile, I saw them only once a month when they came up in the community truck. They never failed to come supplied with "eats" for me. I enjoyed the outdoor oven bread, dried meat, and tamales they usually brought. It took a while to get accustomed to the diet of the school. I longed for my grandmother and my younger brothers and sisters. I longed for my house. I longed to take part in a Buffalo Dance. I longed to be free.

I came home for the 4-day Thanksgiving break. At first, home did not feel right anymore. It was much too small and stuffy. The lack of running water and bathroom facilities were too inconvenient. Everything got dusty so quickly and hardly anyone spoke English. I did not realize I was beginning to take on the white man's ways, the ways that belittled my own. However, it did not take long to "get back with it." Once I established my relationships with family, relatives, and friends I knew I was where I came from and where I belonged.

Leaving for the boarding school the following Sunday evening was one of the saddest events in my entire life. Although I enjoyed myself immensely the last few days, I realized then that life would never be the same again. I could not turn back the time just as I could not do away with school and the ways of the white man. They were here to stay and would creep more and more into my life. The effort to make sense of both worlds together was painful and I had no choice but to do so. The schools, television, automobiles, and other white man's ways and values had chipped away at the simple cooperative life I grew up in. The people of Cochiti were changing. The winter evening gatherings, exchanging of stories, and even the performing of certain ceremonies were already only a memory that someone commented about now and then. Still the demands of both worlds were there. The white man's was flashy, less personal, but comfortable. The Indian was both attracted and pushed toward these new ways that he had little to say about. There was no choice left but to compete with the white man on his terms for survival. For that I knew I had to give up a part of my life.

Determined not to cry, I left for school that dreadfully lonely night. My right hand clutched tightly the mound of cornmeal grandmother placed there and my left hand brushed away a tear as I made my way back to school.

Author Index

Subject Index

A

Abstraction, 81
Academic achievement
  factors affecting, 31
  measurement of, 107
Academic performance
  differences in, 116–117
  poor, explanation of, 101–102
Access, 1–2, 117–118
  equal, 2
Activities, structure of, 92
Addition, conversion into, 171–172
Additive bilingualism, 217
Age, and math ability, 38
Age Differentiation Hypothesis, 39
Algebra, learning of
  by ninth-grade Hispanics, 203–211
  in United States, 196–197
Algebraic word problems, solving of, 30
Applied research project on language proficiency, related to mathematics achievement, 223–224
Arabic counting system, 85
Arithmetic, 192
  and developmental psychology, 75–87

Asian-American students, math achievement, 41, 123–125
  factors affecting, 127–129
  mathematics characteristics, 131–134
Aspirations, related to achievement, 268, 269
Assistance, parental, 36
Attainment, student, 188

B

Background measures, as predictors of reasoning and reading factors, 252–253
Base structure conventions, children's understanding of, 56–58
Behaviors
  student, 108–109
  teacher/aide, 108
  whole class, 109
Belgium, eighth-grade mathematics achievement, 193
Bilingual Education Act, 103
Bilingual programs, 217
  content of, 104
Bilingual student